SpringerBriefs in Mathemat

Series editors

Krishnaswami Alladi
Nicola Bellomo
Michele Benzi
Tatsien Li
Matthias Neufang
Otmar Scherzer
Dierk Schleicher
Vladas Sidoravicius
Benjamin Steinberg
Yuri Tschinkel
Loring W. Tu
G. George Yin
Ping Zhang

SpringerBriefs in Mathematics showcases expositions in all areas of mathematics and applied mathematics. Manuscripts presenting new results or a single new result in a classical field, new field, or an emerging topic, applications, or bridges between new results and already published works, are encouraged. The series is intended for mathematicians and applied mathematicians.

For further volumes:
http://www.springer.com/series/10030

Marco Bramanti

An Invitation to Hypoelliptic Operators and Hörmander's Vector Fields

 Springer

Marco Bramanti
Dipartimento di Matematica
Politecnico di Milano
Milan
Italy

ISSN 2191-8198   ISSN 2191 8201 (electronic)
ISBN 978-3-319-02086-0   ISBN 978-3-319-02087-7   (eBook)
DOI 10.1007/978-3-319-02087-7
Springer Cham Heidelberg New York Dordrecht London

Library of Congress Control Number: 2013949251

Printed on acid-free paper

Springer is part of Springer Science+Business Media (www.springer.com)

Preface

Hörmander's operators are an important class of linear elliptic-parabolic degenerate partial differential operators with smooth coefficients, which have been intensively studied since the late 1960s and are still an active field of research. This text is based on the notes that I wrote for a short course held at the Department of Mathematics of *Politecnico di Milano* in February–March 2012. The audience consisted of a group of colleagues and Ph.D. students working on PDEs. In accordance with the style of that course, this text is written for people who are interested in beginning to do research in this area, or are just curious about it, and look for a *general overview* of this field of research, with its motivations and problems, some of its fundamental results, and at least some of its recent lines of development. In this text *proofs* are almost completely absent, which is an uncommon feature for a mathematical booklet, and therefore needs some justification, which will appear from the rest of this introduction.

Let me first sketch the plan of these notes. Chapters 1 and 2 deal, in two different ways, with *motivations* for studying Hörmander's operators. (Actually, the quest of motivation is a leitmotiv of these notes). Chapter 1 explains the PDE context at the origin of the concept of Hörmander's operators, while Chap. 2 focuses on the relevance in other areas of pure or applied mathematics of specific classes of partial differential operators which are actually of Hörmander's type. Chapters 3 and 4 discuss some fundamental ideas and results in this area, dating back to the 1970s and 1980s, which everyone interested in doing research, or studying contemporary research papers about Hörmander's operators, needs to know. More specifically, Chap. 3, perhaps the technical core of this text, deals with the theme of a-priori estimates, in the suitable Sobolev spaces, for Hörmander's operators. This involves the concept of homogeneous groups, the construction of fundamental solutions, the use of abstract singular integral theories, and the development of suitable algebraic and differential geometric tools. Chapter 4 deals with the geometry of the vector fields which appear in the definition of Hörmander's operators, the concept of distance induced by a system of vector fields, and related problems. Actually, the study of *geometry of Hörmander's vector fields* is nowadays an independent field of research with respect to the study of *second order PDEs of Hörmander's type*. These notes are mainly focussed on the second theme, with a touch on the first one mainly as a tool for the study of the second one. However, in Chap. 4 I have at least tried to give some of the motivations for the study of the geometry of Hörmander's vector fields, besides its applications to PDEs. Finally, Chap. 5 presents an overview of some of the developments of the theory of Hörmander's operators in the 1990s and 2000s. Here the choice of the topics particularly reflects my personal interests. As we will see, the evolution of this area has mainly consisted of the study of classes of operators which are no longer of Hörmander type, strictly speaking, but are structured on Hörmander's vector fields, in several senses, and also contain some nonsmooth ingredients, which poses new problems. Again, I have always tried to give some motivations for the study of the particular classes of operators which are considered.

Let me now try to justify the style of this text. Motivation for the study of Hörmander's operators and vector fields involve, among other areas, systems of stochastic differential equations, the theory of functions of several complex variables, geometric control theory and nonholonomic mechanics: a wide range of subjects, which is impossible to give account of in a few pages. The fundamental ideas and results which still now constitute the basic tools to work on Hörmander's operators appeared in the literature in a small number of very important papers, which are very technical, long, and often not easy to read. A detailed exposition of the contents of those papers, in the style of a graduate course, would require several hundreds of pages. I think that both for a person who is just curious about this area and for one who wants to begin doing active research in it, acquiring from the very beginning a general picture of the landscape can be of great help. Doing this in a limited time requires avoiding proofs. The study of proofs and techniques is a, clearly unavoidable, second step which a person will take, starting with the specific techniques involved in the specific problem he/she wants to attack.

Any comment on this book will be appreciated. Please, write to: marco.bramanti@polimi.it

Acknowledgments

I wish to thank all the people who attended the short course at the Department of Mathematics of Politecnico di Milano for their interest and for stimulating participation to the lessons. I am also grateful to some collegues who read this text and made useful comments and suggestions: Marco Peloso, Ermanno Lanconelli, and Sergio Polidoro.

Contents

Contents

Chapter 1
Hörmander's Operators: What they are

1.1 The Context of Distribution Theory

In 1950 Laurent Schwartz was awarded with the Fields Medal for his creation of the *theory of distributions* (see [16, 17]). Since every distribution is infinitely differentiable but, on the other hand, only the product of a *smooth* function with a distribution is generally a distribution, this theory is a natural framework for the study of *linear partial differential equations with smooth coefficients*. For this class of equations the theory allows to give a very general definition of solution, which turns out to be appropriate under several regards. For instance, in spite of the apparent weakness of the concept of distributional solution, this notion is capable of distinguishing between a solution to $Lu = 0$ and a fundamental solution of the operator L, a distinction which the notion of "equality almost everywhere" fails to reveal. Distribution theory provided for the following decades the conceptual framework for the study of general properties of partial differential operators with smooth coefficients.[1] One of the leaders in this field of research was Lars Hörmander, who won the Fields Medal in 1962 for his deep study of linear PDEs with smooth coefficients. The book [7] represents an account of Hörmander's results in this field up to that time.

Let us consider a linear differential operator of order m, with (real or complex) coefficients, defined and infinitely differentiable on the whole $\mathbb{R}^n$ or in some domain. If we want to establish some general properties, holding independently of the kind of equation (elliptic, hyperbolic. . .), we cannot study a boundary or initial value problem. Instead, two basic questions are:

Does equation $Lu = f$ possess any solution?

If $Lu = f$ and f is smooth, is u smooth?

The first question introduces the notion of *solvability*, the second one that of *hypoellipticity*.

---

[1] An interesting account of the early impact of the theory of distributions on the mathematical environment is given by Lars Gårding in [6, Chap. 12].

M. Bramanti, *An Invitation to Hypoelliptic Operators and Hörmander's Vector Fields*,
SpringerBriefs in Mathematics, DOI: 10.1007/978-3-319-02087-7_1,
© The Author(s) 2014

1.2 Local Solvability

Definition 1 *(See [5, p. 85]) An operator L is said locally solvable at x_0 if there exists a neighborhood $\mathcal{U}$ of x_0 such that for any $f \in C_0^\infty (\mathcal{U})$, equation $Lu = 0$ has a (distributional) solution $u \in \mathcal{D}' (\mathcal{U})$.*

Recall that the fundamental solution $\Gamma_y (\cdot)$ of an operator L with pole y is, by definition, a distributional solution to the equation $L\Gamma_y (\cdot) = \delta_y (\cdot)$ where δ_y is the Dirac mass concentrated at y. If L has constant coefficients then $\Gamma_y (x) = \Gamma_0 (y - x)$ and the equation $Lu = f$ with f compactly supported distribution is solved (at least) by $u = \Gamma_0 * f$. One of the early successes of distribution theory was the following:

Theorem 2 *(Malgrange-Ehrenpreis, 1956) Every linear differential operator with constant complex coefficients has a (distributional) fundamental solution.*

Therefore, *all* linear differential operators *with constant coefficients* are locally solvable.

For operators with variable coefficients the situation is not so good. In 1957, just one year after Malgrange-Ehrenpreis' positive result, H. Lewy [13] found the first example of a linear equation with polynomial coefficients which may not have any solution: namely, the equation

$$\partial_{x_1} u + i \partial_{x_2} u - 2i (x_1 + i x_2) \partial_{x_3} u = f (x_1, x_2, x_3)$$

does not have any solution in any nonempty open set of $\mathbb{R}^3$, for some $f \in C^\infty (\mathbb{R}^3)$. Lewy's example was really striking for the time.[2] Through the years more examples of this kind were found. An interesting one is given by Mizohata's equation (quoted in [5, p. 84]):

$$\partial_x u + i x^k \partial_y u = f$$

which is solvable for k even, but not solvable for k odd (Grushin, 1971). An example of nonsolvable operator with real coefficients is the following in 3 variables (x, y, z), due to Hörmander (see [5, p. 85]):

$$Lu = \left(y^2 - z^2\right) u_{xx} + \left(1 + x^2\right) \left(u_{yy} - u_{zz}\right) - xy\, u_{xy} - (xy\, u)_{xy} + xz\, u_{xz} + (xz\, u)_{xz}.$$

A much easier and surprising example is the following, by Kannai, 1971 [9]: the operator

$$L = x \partial_{yy}^2 + \partial_x$$

is unsolvable at any point $(0, y_0)$; by contrast, the operator

---

[2] Lewy's paper appeared on "Annals of Math.". In its review on the Math. Rev. we read: "Experience with linear partial differential equations has shown that they generally possess smooth local solutions provided the equations are sufficiently smooth. This paper produces the first example of a system with coefficients in C^∞ having no smooth solutions in any domain".

$$L = x\partial_{yy}^2 - \partial_x$$

is locally solvable at any point! The last example also shows that lower order terms are important for local solvability.

Let us now recall the definition of elliptic operator:

Definition 3 *A linear differential operator of order m (with complex coefficients)*

$$L = \sum_{|\alpha| \leq m} a_\alpha(x)\, \partial_x^\alpha$$

is said elliptic if, given its principal symbol

$$p_0(x, \xi) = \sum_{|\alpha| = m} a_\alpha(x)\, \xi^\alpha$$

one has

$$p_0(x, \xi) = 0 \text{ if and only if } \xi = 0.$$

Then, we can state the following:

Theorem 4 *(See [12, Chap. 4]) Any elliptic operator with C^∞ coefficients is locally solvable at any point.*

In [5, pp. 85–90], several necessary or sufficient conditions for solvability are discussed. A criterion by Nirenberg-Treves gives a necessary and sufficient condition for the local solvability of the so-called *operators of principal type* ([5, pp. 85–90]). The definitive result in this direction is due to Beals-Fefferman, 1973 [1], but here we will not go into further details.

Let us now turn to the concept of *hypoelliptic operator* which, as we will see, is also related to that of solvable operator.

1.3 Hypoellipticity

Definition 5 *A differential operator L with $C^\infty(\Omega)$ coefficients (Ω open subset of $\mathbb{R}^n$) is said hypoelliptic in Ω if, for any open set $\Omega' \subset \Omega$ and any distribution $u \in D'(\Omega')$, $Lu \in C^\infty(\Omega') \Rightarrow u \in C^\infty(\Omega')$.*

Analogously, an operator is said analytic-hypoelliptic if $Lu \in C^\omega(\Omega) \Rightarrow u \in C^\omega(\Omega)$ ($C^\omega(\Omega)$ means analytic in Ω).[3]

---

[3] Analogously, one can introduce the notion of Gevray-hypoellipticity, replacing C^ω with Gevray classes. See [5, pp. 91–92].

1.3.1 Hypoelliptic Operators with Constant Coefficients

It follows immediately from the definition that if a hypoelliptic operator with constant coefficients possesses a fundamental solution Γ_0, then $\Gamma_0 \in C^\infty (\mathbb{R}^n \setminus \{0\})$ (analogously, if the operator is analytic-hypoelliptic, any fundamental solution will be analytic outside the origin). A classical theorem by Schwartz states that the converse is true, too:

Theorem 6 *(See [18, Chap. 2, Theorem 2.1]) If the operator L with constant coefficients has a fundamental solution $C^\infty (\mathbb{R}^n \setminus \{0\})$ (or $C^\omega (\mathbb{R}^n \setminus \{0\})$), then L is hypoelliptic (respectively: analytic-hypoelliptic).*

For instance, in two variables we have:

Hypoelliptic and analytic hypoelliptic	Hypoelliptic but not analytic hypoelliptic	Not hypoelliptic
Laplacian: $\partial_{xx}^2 + \partial_{yy}^2$	Heat[4]: $\partial_t - \partial_{xx}^2$	Wave: $\partial_{xx}^2 - \partial_{yy}^2$
Cauchy-Riemann: $\partial_x + i\partial_y$		Schrödinger: $i\partial_t + \partial_{xx}^2$

Within the class of linear differential operators of any order m of *constant (complex) coefficients*, hypoelliptic operators have been characterized by Hörmander. The result is the following:

Theorem 7 *(See [4, p. 80]) Let*

$$L = \sum_{|\alpha| \leq m} a_\alpha \partial_x^\alpha$$

(with a_α complex constants, m positive integer) and let $p(\xi)$ be the polynomial defined, via Fourier transform, by the relation

$$\widehat{Lu}(\xi) = p(\xi)\widehat{u}(\xi).$$

Then L is hypoelliptic if and only if

$$\lim_{|\xi| \to \infty} \frac{|\nabla p(\xi)|}{|p(\xi)|} = 0. \tag{1.1}$$

Example 8 *This criterion allows to check that Laplace equation, the heat equation, Cauchy-Riemann equation are hypoelliptic while the wave equation and Schrödinger equation are not. Less obvious examples are the following, which the reader can check by condition (1.1):*

---

[4] Think to the fundamental solution of the heat equation, which vanishes identically for $t < 0$ but not for $t > 0$, hence is not analytic.

$L = \partial^4_{xxxx} - \partial^2_{yy}$ is hypoelliptic in $\mathbb{R}^2$ while
$L = \partial^4_{xxxx} + \partial^2_{yy}$ is not;
$L = \partial^4_{xxxx} \pm i\partial^2_{yy}$ are hypoelliptic in $\mathbb{R}^2$.

Although condition (1.1) is easy to write, in the class of operators of any order m and with complex constant coefficients, the characterization of hypoelliptic operators is not so transparent, as one sees from the last examples. If we restrict our attention to *second order* operators with *real constant* coefficients the situation is much simpler: hypoelliptic operators are exactly *elliptic* and *parabolic* operators. This fact will follow easily by an important theorem that we will discuss in Sect. 1.3.4.

Also, in the class of operators of any order m with constant complex coefficients, *analytic hypoelliptic* operators are exactly *elliptic* operators. This is the content of a famous theorem by Petrowsky, 1939.

1.3.2 Hypoelliptic Operators with Variable Coefficients

Let us now come to operators with variable coefficients. Here the characterization of hypoelliptic operators is a much more difficult (and still open) problem. First, let us point out the relation between hypoellipticity and local solvability:

Theorem 9 *(See [15, Chap. 3, p. 3]) Let L be a hypoelliptic operator in Ω. Then its transposed L^* is locally solvable at any point of Ω.*

Recall that the transposed L^* of an operator L with C^∞ coefficients is defined by

$$\langle \phi, L^* u \rangle = \langle L\phi, u \rangle \text{ for any } \phi \in \mathcal{D}(\Omega), u \in \mathcal{D}'(\Omega).$$

Since the transposed of a hypoelliptic operator is, in many cases, still hypoelliptic (or is quite the same operator), hypoelliptic operators are "quite often" local solvable. Anyhow, there exist examples of hypoelliptic operators which are not solvable and viceversa. The example by Kannai already quoted is appropriate: the operator

$$L = x\partial^2_{yy} + \partial_x$$

is hypoelliptic but unsolvable at any point $(0, y_0)$; by contrast, the operator

$$L = x\partial^2_{yy} - \partial_x$$

is locally solvable at any point but is not hypoelliptic.

Next, let us point out that elliptic operators are good also under this respect, namely:

Theorem 10 *(See [12, Chap. 4]) Any elliptic operator with C^∞ (or analytic) coefficients is hypoelliptic (respectively, analytic-hypoelliptic).*

As already said, no characterization of hypoelliptic operators with variable coefficients exists so far. The book [7] by Hörmander, 1963, contains a plenty of results giving sufficient conditions in order for an operator to be hypoelliptic. For instance, a first class of operators with variable coefficients which has been studied is that of operators "with constant strength" (a notion introduced by Peetre, 1959), which can be seen as small local perturbations of operators with constant coefficients. (Here we do not go into details. See [5, p. 94] for a survey of results in this direction). For operators of constant strength with continuous coefficients, a local solvability result holds (see [7, Theorem 7.2.1, p. 172]). Moreover, if the operator has C^∞ coefficients and constant strength, hypoellipticity of the frozen operator $P(x_0, D)$ implies hypoellipticity of $P(x, D)$.

1.3.3 An Unsatisfactory Situation

Nevertheless, all the results proved through the years up to 1967 did not enlighten some specific examples of known hypoelliptic operators which showed a degenerate behavior. For instance, already in 1934, Kolmogorov [11] showed that the degenerate parabolic operator

$$L = \partial_t + x\partial_y - \partial_{xx}^2 \tag{1.2}$$

has a fundamental solution smooth (in the joint variables) outside the pole

$$\Gamma\left(t, x, y, t', x', y'\right) = \begin{cases} \dfrac{2\sqrt{3}}{\pi (t-t')^2} \exp\left(-\dfrac{(x-x')^2}{4(t-t')} - \dfrac{3\left(y-y'-\frac{x+x'}{2}(t-t')\right)^2}{(t-t')^3}\right) & \text{for } t > t' \\ 0 & \text{for } t < t' \end{cases},$$

hence it is actually hypoelliptic. On the other hand, if we freeze its coefficients at $(0, y_0, t_0)$ we find an operator acting on two variables only, which therefore cannot be hypoelliptic. This means that the theory of "operators with constant strength" does not explain why Kolmogorov' operator (1.2) is hypoelliptic. This example was not isolated, since in the papers of Weber 1951, Il'in 1964, more general classes of degenerate parabolic operators with a fundamental solution smooth outside the pole were exhibited. All these were examples of equations of Kolmogorov-Fokker-Planck type, which the transition probability density of a stochastic process must obey to, equations having a physical meaning and interest. We will say more on this kind of equations in Sect. 2.1.

Summarizing: *using the existing theories of the early sixties there was no way to decide the hypoellipticity of an operator just looking at the coefficients*; it was sometimes necessary to compute the fundamental solution, a harder task.

1.3.4 A Turning Point: Hörmander 1967, Acta Mathematica

The first result which explained the above phenomena in the framework of a general theory, allowing to predict hypoellipticity of an operator just looking at its coefficients was Hörmander's Theorem of 1967, see [8]. Here the theory concentrates on *second order operators with smooth real coefficients*.

In order to understand the meaning of the main assumption in this result, let us state some preliminary steps. First Hörmander proves that:

Theorem 11 *If a second order differential operator with real C^∞ coefficients is hypoelliptic, then its quadratic form is semidefinite at every point.*

Note that the sign of the quadratic form can change from a point to another, like in the already quoted example by Kannai:

$$L = x\partial_{yy}^2 + \partial_x.$$

Let us consider, therefore, a second order operator with real smooth coefficients and semidefinite quadratic form:

$$L = \sum_{i,j=1}^n a_{ij}(x)\,\partial_{x_i x_j}^2 + \sum_{i=1}^n b_i(x)\,\partial_{x_i} + b_0(x)$$

where, $\forall x \in \Omega$, either

$$a_{ij}(x)\,\xi_i\xi_j \geq 0 \ \forall \xi \in \mathbb{R}^n \text{ or } a_{ij}(x)\,\xi_i\xi_j \leq 0 \ \forall \xi \in \mathbb{R}^n.$$

In any open set where the rank of the matrix $a_{ij}(x)$ is constant, the operator L (or $-L$) can be written in the form:

$$L = \sum_{i=1}^r X_i^2 + X_0 + c$$

where

$$X_i = \sum_{k=1}^n b_{ik}(x)\,\partial_{x_k}$$

with b_{ik}, c smooth on the open set, and the symbol X_i^2 denotes the composition $X_i(X_i)$. Henceforth, we will identify the first order differential operator $X_i = \sum_{k=1}^n b_{ik}(x)\,\partial_{x_k}$ with the vector field

$$X_i = (b_{i1}(x), b_{i2}(x), \ldots, b_{in}(x)).$$

We define the commutator (of step 2) of the vector fields X_i, X_j as the vector field

$$[X_i, X_j] = X_i X_j - X_j X_i$$

(where the product $X_i X_j$ is meant as composition of differential operators, and the result is interpreted as a vector field). Analogously we can define the commutators of step 3,

$$[[X_i, X_j], X_k]$$

and, inductively, of any step r.

Now, let us consider the Lie algebra generated by the vector fields X_i, that is the set of vector fields obtained as linear combination of the fields X_i and their commutators of any finite step. Assume that in some open set this algebra has dimension $m < n$. Then, by a known theorem by Frobenius, there exists, locally, a new coordinate system $y_1, \ldots, y_n$ in which the operator L only depends on the first m variables, and therefore cannot be hypoelliptic.[5] If we want to prove hypoellipticity of L, therefore, it is suggested to assume that the Lie algebra generated by the vector fields X_i span $\mathbb{R}^n$ at every point of the domain. This is exactly "Hörmander's condition". We can now state the following:

Theorem 12 ("Hörmander's theorem") *Let*

$$L = \sum_{i=1}^{q} X_i^2 + X_0 + c \tag{1.3}$$

where the vector fields have real C^∞ coefficients. Assume that the Lie algebra generated by the fields X_i has dimension n at every point of an open set $\Omega \subset \mathbb{R}^n$. Then L is hypoelliptic in Ω.

A bit of terminology: the assumption about the Lie algebra generated by the X_i's is generally known, after this theorem, as "Hörmander's condition". A system of smooth vector fields satisfying Hörmander's condition in some domain Ω is called "a system of Hörmander's vector fields". An operator L like (1.3), where the X_i's are a system of Hörmander's vector fields is called "a (complete) Hörmander's operator"; if $X_0 = 0$ it is called "a sum of squares of Hörmander's vector fields".

Examples. This theorem, in particular, implies the hypoellipticity of those highly degenerate parabolic operators for which fundamental solutions (smooth outside the pole) had been computed (Kolmogorov, Weber, Ilin, etc.). For instance, if we consider, in $\mathbb{R}^3$, the fields

---

[5] To be more precise, the statement "L cannot be hypoelliptic" is correct if the operator, after the change of variables, does not possess a zero order term, for in that case any rough function of the variables which are not involved in the derivatives is a solution to the equation. In the general case one can say that *if the equation $Lu = 0$ admits a nontrivial solution u,* then one can modify u in a suitable halfspace (normal to one of the direction of the variables not involved in the equation), getting a discontinuous function which is still a distributional solution to the equation $Lu = 0$. Hence L is not hypoelliptic.

$$X_1 = \partial_x; \; X_0 = \partial_t + x\partial_y$$

then

$$[X_1, X_0] = \partial_y$$

and the three vector fields $X_1, X_0, [X_1, X_0]$ span $\mathbb{R}^3$ at every point of the space; therefore the *Kolmogorov operator*

$$L = X_1^2 + X_0 = \partial_{xx}^2 + \partial_t + x\partial_y$$

is hypoelliptic.

Let us give some more examples. *Grushin operators*:

$$\partial_{xx}^2 + x\partial_y \text{ in } \mathbb{R}^2, \text{ or } \partial_{xx}^2 + x^2\partial_{yy}^2$$

are hypoelliptic in $\mathbb{R}^2$ (take $X_1 = \partial_x$ and $X_0 = x\partial_y$).

The *sublaplacian* in the Heisenberg group[6]:

$$L = X^2 + Y^2 \text{ with}$$
$$X = \partial_y + 2x\partial_t; \; Y = \partial_x - 2y\partial_t$$

is hypoelliptic, since X, Y and $[X, Y] = -4\partial_t$ span $\mathbb{R}^3$ at every point.

Consider also an ultraparabolic, strongly degenerate operator like:

$$L = \partial_{x_1 x_1}^2 + x_1\partial_{x_2} + x_2\partial_{x_3} + \dots + x_{n-1}\partial_{x_n} - \partial_t \text{ in } \mathbb{R}^{n+1}.$$

L is hypoelliptic because, letting

$$X_1 = \partial_{x_1}; \; X_0 = x_1\partial_{x_2} + x_2\partial_{x_3} + \dots + x_{n-1}\partial_{x_n} - \partial_t$$

we have

$$[X_1, X_0] = \partial_{x_2}$$
$$[X_1, [X_1, X_0]] = \partial_{x_3}$$
$$\dots$$
$$[X_1, [X_1, [\dots [X_1, X_0]]]] = \partial_{x_n}$$

so that the commutators up to length n span $\mathbb{R}^{n+1}$ at any point.

Remark 13 (**Operators with constant coefficients**) *If the operator L is hypoelliptic and has constant coefficients, then by the above arguments L (or −L) has*

---

[6] We will explain what is the Heisenberg group in Sect. 2.2.4. For the moment, just consider this operator as defined in $\mathbb{R}^3$.

nonnegative characteristic form and can be rewritten in the form (1.3) in $\mathbb{R}^n$. By the above argument based on Frobenius theorem, since for operators with constant coefficients Hörmander's condition holds everywhere or fails everywhere, the condition in this case is also necessary for hypoellipticity. But for constant vector fields all the commutators vanish, hence Hörmander's condition holds if and only if either $X_1, X_2, ..., X_q$ span $\mathbb{R}^n$, and in this case $q \geq n$ and the operator is elliptic, or $X_0, X_1, X_2, ..., X_q$ span $\mathbb{R}^n$ (but the same is not true without X_0) and in this case $q \geq n - 1$ and the operator is parabolic. We conclude that the only second order hypoelliptic operators with real constant coefficients are elliptic and parabolic operators.

Remark 14 (Transposed of Hörmander's operators and solvability) *The transposed of a vector field*

$$X_i = \sum_{j=1}^n b_{ij}(x)\,\partial_{x_j} \text{ is } X_i^* = -\sum_{j=1}^n \partial_{x_j}\left(b_{ij}(x)\cdot\right) = -X_i + f_i(x),$$

with $f_i = -\sum_{j=1}^n \left(\partial_{x_j} b_{ij}\right)$. Hence the transposed of a Hörmander's operator L like (1.3) is

$$L^* = \sum_{i=1}^q \left(X_i^*\right)^2 + X_0^* + c$$

$$= \sum_{i=1}^q (-X_i + f_i)(-X_i + f_i) - X_0 - f_0 + c$$

$$= \sum_{i=1}^q X_i^2 - X_0 + c' - \sum_{i=1}^q f_i X_i$$

$$= \sum_{i=1}^q X_i^2 + X_0' + c'$$

with

$$X_0' = -X_0 - \sum_{i=1}^q f_i X_i, \quad \text{where } f_i \text{ are smooth functions.} \tag{1.4}$$

Now, it is easy to check that
If $X_0, X_1, ..., X_q$ satisfy Hörmander's condition in Ω and X_0' is like in (1.4), then also $X_0', X_1, ..., X_q$ satisfy Hörmander's condition in Ω.
Namely,

$$[X_k, X_0'] = -[X_k, X_0] - \sum_{i=1}^q f_i [X_k, X_i] - \sum_{i=1}^q (X_k f_i) X_i$$

hence, at any point of Ω, *any commutator* $[X_k, X_0]$ *can be written as a linear combination of the commutators* $\left[X_k, X_0'\right]$, $[X_k, X_i]$ *and the vector fields* X_i. *Iterating this reasoning we see that the Lie algebra generated by* $X_0', X_1, ..., X_q$ *coincides with that generated by* $X_0, X_1, ..., X_q$. *We conclude that:*

Proposition 15 *The transposed* L^* *of a Hörmander's operator* L *is still a Hörmander's operator; in particular,* L^**is hypoelliptic, by Hörmander's theorem. Hence by Theorem 6 the operator* $(L^*)^* = L$ *is solvable. In conclusion:* all Hörmander's operators are locally solvable.

Remark 16 (Properties of the class of Hörmander's operators) *We can also summarize the previous reasoning saying that: the class of Hörmander's operators is closed under transposition, addition of lower order terms (that is linear combinations of the* X_i *for* $i = 1, 2, ..., q$), *changing of sign of the drift* X_0; *if the drift* X_0 *is not required to satisfy Hörmander's condition, we can also add any drift* X_0'. *Hörmander's operators are both hypoelliptic and locally solvable.*

Remark 17 (Some limits of Hörmander's theorem) *In the following we will almost exclusively deal with Hörmander's operators, which are both hypoelliptic and solvable. Before going on, let us stress what Hörmander's theorem does not cover, and in which sense the sufficient condition for hypoellipticity given by Hörmander's theorem is not necessary:*

1. the theorem only considers 2nd order operators;

2. the theorem only considers operators with real coefficients. For operators of type (1.3) with complex coefficients, Hörmander's condition is not sufficient for hypoellipticity. For instance, we have already quoted the sublaplacian

$$L = \left(\partial_y + 2x\partial_t\right)^2 + (\partial_x - 2y\partial_t)^2$$

which is hypoelliptic by Hörmander's theorem; it can be proved that, in contrast with this, the operator

$$L = X_1^2 + X_2^2 + X_0 = \left(\partial_y + 2x\partial_t\right)^2 + (\partial_x - 2y\partial_t)^2 - 4i\,\partial_t$$

is not hypoelliptic, even though Hörmander's condition is fulfilled.[7]

3. The theorem only considers operators which are already written in the form (1.3). But not every 2nd order operator with semidefinite quadratic form and smooth coefficients can be written globally in this form (with smooth vector fields). The example by Kannai, already quoted, is still appropriate:

$$L = x\partial_{yy}^2 + \partial_x$$

is not a Hörmander operator (because we should define $X_1 = \sqrt{x}\partial_y$, *and this vector field is neither smooth nor real valued). Recall that* L *is hypoelliptic but not solvable,*

---

[7] In Sect. 2.2.4 we will say more about this.

while Hörmander's operators are always solvable. A more involved counterexample, of an operator which is nonnegative in the whole $\mathbb{R}^n$ (without changing the sign of the quadratic form) and cannot be written in the form (1.3) is quoted in [14, p. 8]. A systematic study of the operators with nonnegative quadratic form, not necessarily written in the form of sum of squares, is carried out in the book by Olenik-Radkevic [14], 1973.[8] I suggest the reading of the introduction of [14], where a rich account of the research in this field up to that time is given. The study of general operators with nonnegative characteristic form is a difficult field which, through the years, has not been studied by many authors.

4. Even in the class of the operators that the theorem considers, Hörmander's condition is sufficient but not necessary. The argument based on Frobenius' theorem shows that Hörmander's condition cannot be violated everywhere in an open set, but does not exclude the failure of the condition in isolated points or on manifolds of dimension strictly less than n. For instance, Olenik-Radkevic [14] prove that Hörmander's condition can fail in a finite number of points (provided that not all the vector fields vanish at those points) without losing the hypoellipticity of the operator. Between these two endpoint situations (failure of Hörmander's condition in an open set–failure at some isolated points) there are interesting intermediate cases: Fedii, 1971 (quoted in [3]) has given the following example of operator in $\mathbb{R}^2$:

$$L = \partial^2_{xx} + a(x)^2 \, \partial^2_{yy}$$

where $a \in C^\infty(\mathbb{R})$, even, ≥ 0, nondecreasing in $[0, \infty)$, $a(x) = 0 \Leftrightarrow x = 0$. Under these assumptions, L is hypoelliptic. Note that $L = X^2 + Y^2$ but Hörmander's condition holds only if $a(x)$ vanishes of finite order at $x = 0$. For instance, the operator

$$L = \partial^2_{xx} + e^{-2/x^2}\partial^2_{yy} \tag{1.5}$$

can be proved to be hypoelliptic, but the vector fields $X_1 = \partial_x$, $X_2 = e^{-1/x^2}\partial_y$ do not satisfy Hörmander's condition on the line $x = 0$. M. Christ has systematized several known examples of this kind, giving sufficient conditions for hypoellipticity, for operators with coefficients which vanish of infinite order. See [3] for a survey on this subject, and other papers of the same author. Nevertheless, a simple characterization, or a fairly general necessary condition for hypoellipticity is still unknown, and perhaps cannot exist.

In the example (1.5), the function $a(x)$ is C^∞ but not analytic; this is not a pure chance, namely:

Theorem 18 *If L is an operator sum of squares of analytical vector fields, and L is hypoelliptic, then Hörmander's condition holds.*

---

[8] In particular, the first half of that book is devoted to boundary value problems for these elliptic-parabolic degenerate operators, a topic we will not touch here.

1.3.5 Subelliptic Estimates

An alternative proof of Hörmander's theorem [8] has been given in 1973 by Kohn [10], using the language and techniques of pseudodifferential operators.[9] Apart from being perhaps shorter than Hörmander's proof, this proof is interesting in its own for the intermediate result it contains, namely the so-called *subelliptic estimates* in fractional Sobolev spaces of small positive order:

$$\|u\|_{H^{\varepsilon,2}} \le c \left(\|Lu\|_{L^2} + \|u\|_{L^2} \right) \tag{1.6}$$

for any $u \in C_0^\infty(\mathbb{R}^n)$, some $\varepsilon > 0$. Namely, at least in the case of a "sum of squares" operator, one can take $\varepsilon = 1/k$ where k is highest order of commutators required to fulfil Hörmander's condition. Once this basic estimate is proved, it can be iterated to get

$$\|u\|_{H^{s+\varepsilon,2}} \le c_{m,s} \left(\|Lu\|_{H^{s,2}} + \|u\|_{H^{-m,2}} \right) \tag{1.7}$$

for any $s, m > 0$. This (more precisely, a localized version of it which makes use of cutoff functions) implies hypoellipticity. Estimates (1.7) are called "subelliptic estimates" and operators satisfying them "subelliptic operators". Subelliptic estimates are useful in several other contexts, since they express the regularizing properties of L in a quantitative way. As Kohn points out in [10], the possibility of deriving (1.7) from (1.6) depends on the structure of the operator: for instance, the wave operator satisfies (1.6) but not (1.7). Remarkably, the operator (1.5), which is hypoelliptic but does not satisfy Hörmander's condition, does not satisfy subelliptic estimates (see [2, p. 204]).

References

1. Beals, R., Fefferman, C.: On local solvability of linear partial differential equations. Ann. Math. **97**(2), 482–498 (1973)
2. Chen, S.C., Shaw, M.C.: Partial differential equations in several complex variables. AMS/IP Studies in Advanced Mathematics, vol. 19. American Mathematical Society, Providence, RI; International Press, Boston, MA (2001)
3. Christ, M.: Hypoellipticity in the infinitely degenerate regime. In: McNeal J.D. (ed.) Complex Analysis and Geometry (Columbus, OH, 1999), Ohio State University Mathematical Research Institute Publications, vol. 9, pp. 59–84. de Gruyter, Berlin (2001)
4. Egorov, Y.V., Shubin, M.A.: Partial Differential Equations. I: Foundations of the Classical Theory of Partial Differential Equations. Encyclopaedia of Mathematical Sciences, vol. 30. Springer, Berlin (1992)
5. Egorov, Y.V.: Microlocal analysis. In: Egorov, Y.V., Shubin, M.A. (eds.) Partial Differential Equations IV. Encyclopaedia of Mathematical Sciences, vol. 33. Springer, Berlin (1993)

[9] Besides the original paper by Kohn [10], more detailed and readable expositions of this proof can be found for instance in: [14, Chap. II, Sect. 5], [2, Sect. 8.2] or [15, Appendix] (this last reference only deals the case $X_0 = 0$).

6. Gårding, L.: Some Points of Analysis and Their History. University Lecture Series, vol. 11. American Mathematical Society, Providence (1997)
7. Hörmander, L.: Linear Partial Differential Operators. Springer, Berlin (1963)
8. Hörmander, L.: Hypoelliptic second order differential equations. Acta Math. **119**, 147–171 (1967)
9. Kannai, Y.: An unsolvable hypoelliptic differential operator. Israel J. Math. **9**, 306–315 (1971)
10. Kohn, J.J.: Pseudo-differential operators and hypoellipticity. In: Partial Differential Equations. Proceedings of Symposia in Pure Mathematics, vol. XXIII, University of California, Berkeley, CA, 1971, pp. 61–69. American Mathematical Society, Providence, RI (1973)
11. Kolmogorov, A.N.: Zufällige Bewegungen (zur Theorie der Brownschen Bewegung). Ann. Math. (2) **35**(1), 116–117 (1934)
12. Krantz, S.G.: Partial Differential Equations and Complex Analysis. Studies in Advanced Mathematics. CRC press, Boca Raton (1992)
13. Lewy, H.: An example of a smooth linear partial differential equation without solution. Ann. Math. (2) **66**, 155–158 (1957)
14. Oleĭnik, O.A., Radkevič, E.V.: Second Order Equations with Nonnegative Characteristic Form. Plenum Press, New York (1973)
15. Ricci, F.: Sub-Laplacians on Nilpotent Lie Groups. Notes for a course held in 2002/2003 at S.N.S. Pisa. Downloadable at: http://homepage.sns.it/fricci/papers/sublaplaciani.pdf
16. Schwartz, L.: Théorie des distributions. Tome I. Actualités Sci. Ind., no. 1091 = Publ. Inst. Math. Univ. Strasbourg 9. Hermann & Cie., Paris (1950)
17. Schwartz, L.: Théorie des distributions. Tome II. Actualités Sci. Ind., no. 1122 = Publ. Inst. Math. Univ. Strasbourg 10. Hermann & Cie., Paris (1951)
18. Treves, F.: Basic Linear Partial Differential Equations. Academic Press, New York (1975)

Chapter 2
Hörmander's Operators: Why they are Studied

We now want to describe some examples of PDEs written in the form of Hörmander's operators which arise both from physical applications and from other fields of mathematics, to give some more motivation for the study of these equations. We will deal with two different areas, which represent the main "historical" motivations for this field of research, namely Kolmogorov-Fokker-Planck equations arising in the study of stochastic systems, and some relevant PDEs which arise in the theory of several complex variables.

We stress the fact that here we confine ourself to the motivations and applications of *Hörmander's operators*; other motivations for the study of *Hörmander's vector fields* and the related geometry (not directly related with second order PDEs) will be discussed in Sect. 4.1.

2.1 First Motivation: Kolmogorov-Fokker-Planck Equations

2.1.1 Brownian Motion and Langevin's Equation

In 1826 the British botanist Robert Brown [5] noticed that small pollen grains suspended in water are subject to a perpetual irregular movement, which can be microscopically observed. Various interpretations of this phenomenon were proposed in the following decades; in 1877, Delsaux (quoted in [8], p. 87) expressed for the first time the idea that this movement is caused by the impacts of the molecules of the liquid on the suspended particles. A rigorous theory on the Brownian movement was developed by Einstein in several papers appeared in the years 1905 to 1908 (collected in [8]); Einstein showed how the Brownian movement can be predicted on the basis of the molecular-kinetic theory of heat. A different approach was proposed in 1908 by Langevin [16], who arrived to similar quantitative predictions introducing a stochastic differential model. In 1923, Wiener [29] introduced what is now called "Wiener process", as a mathematical model of Brownian motion. As we shall

M. Bramanti, *An Invitation to Hypoelliptic Operators and Hörmander's Vector Fields*,   15
SpringerBriefs in Mathematics, DOI: 10.1007/978-3-319-02087-7_2,
© The Author(s) 2014

see, the points of view of Wiener and Langevin significantly differ. The theory of stochastic processes was rigorously founded, mathematically, in the early 1930s by Kolmogorov. After this turning point, the mathematical study of Brownian motion was carried on by many people, starting with Ornstein and Uhlenbeck [20], 1930; Chandrasekhar [6], 1943, and many others. Here we will present Langevin's model (see [26, Sect. 2.1], [6, Chap. 2]).

Consider a small particle suspended in a liquid. The liquid affects the movement of the particle in two ways: the viscosity of the liquid decelerates the particle, while the random molecular collisions can be modeled as a random external force. Newton's equation for the particle of velocity $\mathbf{v}(t)$ and mass m reads as follows (after dividing for m):

$$\mathbf{v}' = -\beta \mathbf{v} + \mathbf{f}(t). \tag{2.1}$$

This differential equation is called Langevin equation; the key point is that the term $\mathbf{f}(t)$ is a random function of t. If we assume that the particle is a homogeneous sphere of radius a, then the coefficient β can be specified according to Stokes' law, as

$$\beta = \frac{6\pi a \eta}{m}$$

where η is the coefficient of dynamic viscosity of the fluid.

Since the motion we are describing is rather chaotic, we can expect the term $\mathbf{f}(t)$ (which has the dimension of an acceleration) to have rough regularity properties. Namely, $\mathbf{f}(t)$ is assumed to be a *Gaussian white noise*, a notion which we are going to discuss briefly.

2.1.2 Wiener Process and Gaussian White Noise

Here we will be rather informal. For a rigorous introduction to stochastic equations, I suggest the book by Schuss [26] or the notes by Barchielli [2]. For a comprehensive discussion of Kolmogorov-Fokker-Planck equations and their applications, see the book by Risken, [25].

The unidimensional Wiener process is a stochastic process $w(t)$ determined by the following conditions:

1. $w(t) \sim N(0, t)$;
2. the increments $w(t + s) - w(t)$ and $w(t) - w(t - u)$ are mutually independent, and are independent of t for $s \geq 0$ and $u \geq 0$;
3. the paths $w(t)$ are continuous;
4. the joint probability distribution of $(w(t_1), w(t_2), \ldots, w(t_n))$ is Gaussian for every finite sequence $t_1 < t_2 < \ldots < t_n$.

We will call n-dimensional Wiener process a vector $\mathbf{w}(t)$ having for components n independent unidimensional Wiener processes.

The n-dimensional Gaussian white noise (which we want to use as a definition of the random force $\mathbf{f}(t)$ appearing in Langevin equation) is, morally speaking, the time derivative of a Wiener process: $\mathbf{f}(t) = \frac{d\mathbf{w}(t)}{dt}$. This is a actually a troublesome definition because it can be proved that *almost surely the trajectory of* $\mathbf{w}(t)$ *is nowhere differentiable*. We then rewrite Eq. (2.1) in terms of differentials

$$d\mathbf{v} = -\beta \mathbf{v} dt + \sqrt{\frac{2\beta kT}{m}} d\mathbf{w}(t), \qquad (2.2)$$

which is the standard form for a stochastic equation. (The exact form of the constant multiplying $d\mathbf{w}$ can be derived by a physical reasoning). Note that the passage from $\mathbf{w}'(t)$ to $d\mathbf{w}(t)$ is also reasonable on a physical ground, since $\mathbf{w}'(t)$ is a force (per unit mass) while $d\mathbf{w}(t)$ is an impulse (per unit mass), which is a more natural quantity to describe a collision. The term $d\mathbf{w}(t)$ appearing in (2.2) is called n-*dimensional Gaussian white noise*. Taking into account the initial condition $\mathbf{v}(0) = \mathbf{v}_0$ we can also integrate the previous equation, writing

$$\mathbf{v}(t) = \mathbf{v}_0 - \int_0^t \beta \mathbf{v}(\tau) d\tau + \sqrt{\frac{2\beta kT}{m}} \mathbf{w}(t). \qquad (2.3)$$

Note that now the Wiener process enters the equation without being differentiated.

Remark 19 *"Mathematical" Brownian motion versus "physical" Brownian motion The Wiener process* $\mathbf{w}(t)$ *is also called (by mathematicians) Brownian motion. This is quite confusing in the particular example we are discussing, because a mathematical model of the physical Brownian motion (according to Langevin) is described by the process* $\mathbf{x}(t)$ *such that* $\mathbf{v}(t) = \mathbf{x}'(t)$ *satisfies (2.3). Now, the process* $\mathbf{x}(t)$ *is actually very different from* $\mathbf{w}(t)$. *For instance,* $\mathbf{w}(t)$ *is never differentiable while* $\mathbf{x}(t)$ *is actually differentiable, because it is the primitive of* $\mathbf{v}(t)$. *To avoid confusion we will always call* $\mathbf{w}(t)$ *"Wiener process".*

2.1.3 Stochastic Differential Equations

Let us now generalize the previous example. Let $w(t)$ be a unidimensional Wiener process. A process $x(t)$ is said to satisfy the *stochastic differential equation*

$$dx(t) = a(x(t), t) dt + b(x(t), t) dw(t); \quad x(0) = x_0 \qquad (2.4)$$

(where a, b are deterministic functions) if the following *Itô integral equation* is satisfied:

$$x(t) = x_0 + \int_0^t a(x(s), s) \, ds + \int_0^t b(x(s), s) \, dw(s). \qquad (2.5)$$

For the precise meaning of the two integrals appearing in (2.5) in the general case, see [26, Chap. 3]. For the moment, it will be enough to observe that:

(1) the first integral in (2.5) represents a random process, if $x(s)$ is a process with continuous paths and $a(x, s)$ is a continuous function;
(2) if the function b does not depend on x (a particular case which happens for instance in the Langevin's Eq. (2.3) and is a differentiable function of time, then the second integral in (2.5) can be defined via integration by parts, as follows:

$$\int_0^t b(s) \, dw(s) = b(t) \, w(t) - \int_0^t w(s) \, b'(s) \, ds$$

where the integrand of the last integral is a continuos random process.

Analogously one can define the meaning of the *system of stochastic o.d.e's*:

$$d\mathbf{x}(t) = \mathbf{b}(\mathbf{x}(t), t) \, dt + \mathbf{B}(\mathbf{x}(t), t) \, d\mathbf{w}(t); \quad \mathbf{x}(0) = \mathbf{x}_0$$

where $\mathbf{w}(t)$ is an n-dimensional Wiener process (we also say: $d\mathbf{w}(t)$ is an n-dimensional white noise), and $\mathbf{b}$, $\mathbf{B}$ are deterministic vector and matrix functions.

2.1.4 Kolmogorov and Fokker-Planck Equations

Let $\mathbf{x}(t)$ be the solution of the stochastic system:

$$d\mathbf{x}(\tau) = \mathbf{b}(\mathbf{x}(\tau), \tau) \, d\tau + \mathbf{B}(\mathbf{x}(\tau), \tau) \, d\mathbf{w}(\tau); \quad \mathbf{x}(t) = \mathbf{x}. \qquad (2.6)$$

Then the *transition probability density* $p(t, x, s, y)$ $(s > t)$, defined by the relation

$$P(\mathbf{x}(s) \in A | \mathbf{x}(t) = \mathbf{x}) = \int_A p(t, \mathbf{x}, s, \mathbf{y}) \, d\mathbf{y}$$

can be proved to satisfy the *forward Kolmogorov equation* (also called Fokker-Planck equation) in the variables (s, y):

$$\partial_s p + \nabla_{\mathbf{y}} \cdot (\mathbf{b}p) - \frac{1}{2} \sum_{i,j=1}^n \partial^2_{y_i y_j} (a_{ij} p) = 0 \qquad (2.7)$$

while $p(t, x, s, y)$ satisfies the *backward Kolmogorov equation* in the variables (t, x):

$$\partial_t p + \mathbf{b} \cdot \nabla_\mathbf{x} p + \frac{1}{2} \sum_{i,j=1}^{n} a_{ij} \partial^2_{x_i x_j} p = 0. \tag{2.8}$$

In (2.7) and (2.8) the matrix $\mathbf{A} = (a_{ij})$ is defined by:

$$\mathbf{A} = \mathbf{B}\mathbf{B}^T$$

with $\mathbf{B}$ as in (2.6). Note that $\mathbf{A}$ *is always symmetric and nonnegative*, but needs not to be positive. Also, note that Eqs. (2.7), (2.8) are *linear partial differential equations*, while (2.6) is a *(generally nonlinear) system of stochastic ordinary differential equations.*

Equations (2.7), (2.8) were first established by Fokker [9] in 1914, Planck [23] in 1917, Kolmogorov [14] in 1931.

Example 20 *Let us come back to the physical Brownian motion of a particle. We want to describe it in the 6-dimensional phase space of variables* $(x_1, x_2, x_3, v_1, v_2, v_3)$. *The system is:*

$$\begin{cases} dx_i = v_i dt & i = 1, 2, 3 \\ dv_i = -\beta v_i dt + \sqrt{\frac{2\beta kT}{m}}\, dw_i & i = 1, 2, 3 \end{cases} \tag{2.9}$$

with dw_i *independent white noises. Then:*

$$\mathbf{b} = (v_1, v_2, v_3, -\beta v_1, -\beta v_2, -\beta v_3)^T\,;$$

$$\mathbf{B} = \sqrt{\frac{2kT}{\beta m}} \begin{pmatrix} 0\,0\,0\,0\,0\,0 \\ 0\,0\,0\,0\,0\,0 \\ 0\,0\,0\,0\,0\,0 \\ 0\,0\,0\,1\,0\,0 \\ 0\,0\,0\,0\,1\,0 \\ 0\,0\,0\,0\,0\,1 \end{pmatrix}; \quad \mathbf{A} = \frac{2\beta kT}{m} \begin{pmatrix} 0\,0 \\ 0\,I \end{pmatrix}$$

and the backward Kolmogorov equation is:

$$\partial_t p + \sum_{i=1}^{3} v_i \partial_{x_i} p - \beta \sum_{i=1}^{3} v_i \partial_{v_i} p + \frac{\beta kT}{m} \sum_{i,j=1}^{3} \partial^2_{v_i v_i} p = 0. \tag{2.10}$$

Note that (2.10) is an ultraparabolic equation: the second order operator is a "partial Laplacian" involving only 3 of the 6 "spacial" variables. The interesting feature of Eq. (2.10), from the analytical point of view, is that, despite its degeneracy, its left hand side is a hypoelliptic Hörmander's operator. Namely, Eq. (2.10) can be written as

$$\sum_{i=1}^{3} X_i^2 p + X_0 p = 0$$

with

$$X_i = \sqrt{\frac{\beta k T}{m}} \partial_{v_i} \ (i = 1, 2, 3) \, ; \ X_0 = \partial_t + \sum_{i=1}^{3} v_i \partial_{x_i} - \beta \sum_{i=1}^{3} v_i \partial_{v_i}$$

and computing the commutators

$$[X_i, X_0] = \sqrt{\frac{\beta k T}{m}} \left(\partial_{x_i} - \beta \partial_{v_i} \right)$$

one sees that the 7 fields X_i, $[X_i, X_0]$, X_0 $(i = 1, 2, 3)$ span $\mathbb{R}^7$ at any point. Then, by Hörmander's Theorem, Eq. (2.10) is hypoelliptic.

Remark 21 *An intuitive interpretation of KFP equations as transport-diffusion equations. The following fact is also worthwhile to be noted. Comparing (2.9) with (2.10) we see that, in the phase space, Eq. (2.10) is degenerate exactly in the directions which do not involve randomness: in system (2.9), the first 3 equations do not contain noise. One can read this fact saying that the probability of finding the system in some phase evolves in time as a diffusion process, diffusing (at least "for short time", that is infinitesimally) in the directions which involve randomness. If the system were totally deterministic, with delta (that is "certain") initial conditions, the solution would be a delta distribution at any time, and Kolmogorov equation would be a first order transport equation. Moreover, if the drift term X_0 had constant coefficients, there would be no interaction between the two phenomena of diffusion and transport, and the probability density would keep diffusing in a lower dimensional subspace of $\mathbb{R}^6$. Instead, the validity of Hörmander's condition (made possible by the fact that the coefficients are not constant) means that, as a second order effect of the interaction between the diffusion matrix and the drift term, diffusion actually takes place in the full space $\mathbb{R}^6$. This diffusion of information in the full space preserves regularity of solutions.*

Remark 22 *Probabilistic proof of Hörmander's theorem. The link between Hörmander's operators and Kolmogorov equations is so relevant from the standpoint of stochastic analysis that a probabilistic proof of Hörmander's theorem has been developed by Malliavin, [17], 1976, using the stochastic calculus of variations, also called Malliavin calculus. In this field, the hypoellipticity of Kolmogorov operator is interpreted as a result of regularity for the probability law of a process. An exposition of Malliavin calculus, with application to the proof of Hörmander's theorem, can be found in the book [19].*

2.1.5 Examples of Kolmogorov-Fokker-Planck Equations Arising from Applications of Stochastic Models

The Brownian harmonic oscillator (See Schuss [26, pp. 45–46]). For an electron in a ionized gas (plasma), the elastic force is the Coulomb force of attraction and repulsion of the neighboring ions. The Langevin equation reads as:

$$x'' + \beta x' + \omega^2 x = w'(t)$$

corresponding to the stochastic system:

$$\begin{cases} dx_1 = x_2 dt \\ dx_2 = -\omega^2 x_1 dt - \beta x_2 dt + dw \end{cases}$$

and the Backward Kolmogorov equation is:

$$\partial_t p + x_2 \partial_{x_1} p - \left(\omega^2 x_1 + \beta x_2 \right) \partial_{x_2} p + \frac{1}{2} \partial^2_{x_2 x_2} p = 0.$$

Letting

$$X_1 = \frac{1}{\sqrt{2}} \partial_{x_2}, X_0 = \partial_t + x_2 \partial_{x_1} - \left(\omega^2 x_1 + \beta x_2 \right) \partial_{x_2}$$

one can check that Hörmander's condition holds. Hence the operator is hypoelliptic, although ultraparabolic.

Brownian motion in periodic potentials (See Risken [25, Chap. 11]). A more general situation is that of a Brownian particle moving in a force field characterized by a periodic potential, plus (possibly) a constant external force. An example of this situation is given by a superionic conductor (see [25, p. 280]), consisting in an almost fixed lattice of ions, in which some other ions are highly movable. The ions forming the lattice generate an electric field, which is periodic in space; but the lattice is also subject to small vibrations, which can be modeled as a white noise perturbation of the periodic field. The lattice also opposes to the movement with a friction force, so the stochastic system is:

$$\begin{cases} dx_i = v_i dt & i = 1, 2, 3 \\ dv_i = -\beta v_i dt + \partial_{x_i} U dt + \sqrt{\frac{2\beta kT}{m}} dw_i & i = 1, 2, 3 \end{cases}$$

where $U(\mathbf{x})$ is the potential. Backward Kolmogorov equation:

$$\partial_t p + \sum_{i=1}^{3} v_i \partial_{x_i} p + \sum_{i=1}^{3} \left(\partial_{x_i} U(\mathbf{x}) - \beta v_i \right) \partial_{v_i} p + \frac{\beta kT}{m} \Delta_v p = 0$$

where Δ_v is the (partial) Laplacian in the three v variables. This ultraparabolic equation is still a hypoelliptic equation of Hörmander type.

Finance, study of "Asian options" (see Barucci-Polidoro-Vespri [3]. For a survey on the general subject of Kolmogorov equations in finance see [22]; a reference book is [21]). Consider a "stock" whose price $S(t)$ at time t satisfies the stochastic equation

$$dS(t) = \mu_0 S(t)\, dt + \sigma S dW$$

where μ_0, σ are constants and dW is a white noise; we are interested in the price V of an *option*, which is the right (without obligation) to buy a stock at a given exercise price E at some fixed "expiry time" T. The price V is a function of t and S (the "underlying" stock). If E is a constant (like in "European options") V satisfies a final condition of the kind

$$V(T, S_T) = \max(S_T - E, 0).$$

In the so-called "Asian options" the exercise price is not a constant, but depends on some average on the history of the stock price S. One of these possibilities is that of *geometric average floating strike call option*; the average is defined as:

$$A(t) = \int_0^t \log S(\tau)\, d\tau.$$

In this case the option price is a function $V(t, S, A)$ and the final condition is

$$V(T, S(T), A(T)) = \max\left(S(T) - e^{A(T)/T}, 0\right).$$

It can be proved that V satisfies in this case the following PDE:

$$-rV + \partial_t V + rS\partial_S V + \frac{1}{2}\sigma^2 S^2 \partial_{SS}^2 V + \log(S)\, \partial_A V = 0$$

(r, σ constants). This equation is rather involved, but with the tricky change of variables

$$x = \frac{\sqrt{2}}{\sigma}\log(S)\,;\ y = \frac{\sqrt{2}}{\sigma}A;$$

$$u(x, y, t) = e^{\frac{2r-\sigma^2}{2\sqrt{2}\sigma}x + \left(\frac{2r-\sigma^2}{2\sqrt{2}\sigma}\right)^2 t}\, V\left(T - t, e^{\frac{\sigma x}{\sqrt{2}}}, \frac{\sigma y}{\sqrt{2}}\right)$$

it is shown to be equivalent to

$$\partial_{xx}^2 u + x\partial_y u - \partial_t u = 0$$

which is exactly the Eq. (1.2) studied by Kolmogorov in 1934 (!), and is of Hörmander type.

Is one always so lucky? Surely not. The same authors, studying Asian options with *arithmetic average floating strike call option*, that is

$$A(t) = \int_0^t S(\tau) \, d\tau$$

with final condition

$$V(T, S(T), A(T)) = \max\left(S(T) - \frac{A(T)}{T}, 0\right)$$

find the PDE

$$-rV + \partial_t V + rS\partial_S V + \frac{1}{2}\sigma^2 S^2 \partial_{SS}^2 V + S\partial_A V = 0.$$

This equation, by a suitable change of coordinates, rewrites as

$$x^2 \partial_{xx}^2 u + x\partial_y u - \partial_t u = 0 \tag{2.11}$$

which is of type $X_1^2 + X_0$ where X_1, X_0 *do not* satisfy Hörmander's condition at $x = 0$.

Computer vision, the Mumford equation or "the process of random direction"
(See Mumford [18]). One of the basic problems in computer vision is to reconstruct the three-dimensional shape and position of real objects starting from some two-dimensional image of them, namely an intensity function $I(x, y)$. The discontinuities of $I(x, y)$ along the "edges" in the image are a central object in this study. D. Mumford is one of the Authors that have used stochastic models for attacking the problem. Quoting from [18, p. 495]:

> What sort of stochastic process is a plausible candidate for modeling the relative likelihood of different edges appearing in a scene of the world? Our edges are to be continuous and almost everywhere differentiable so that, when occluded in part, they will tend to reappear with approximately the same tangent line. The simplest way to do this is to allow curvature $\kappa(s)$, as a function of arc length, to be white noise $w'(s)$, so that once integrated, the tangent direction $\theta(s)$ is a Wiener process[1] $w(s)$. For this reason, we could give this process the name of "process of random direction".

In symbols, we have the stochastic process (x, y, θ) as a function of "time" s (which actually represents arc length) which obeys to the system:

---

[1] In the quotation I have used the symbol w' instead of n used by Mumford, and replaced the term "Brownian motion" with "Wiener process", to keep the terminology we are using. This does not change the meaning of the text.

$$\begin{cases} dx = \cos\theta\,ds \\ dy = \sin\theta\,ds \\ d\theta = \sigma\,dW. \end{cases}$$

The backward Kolmogorov equation is:

$$\partial_t p + \cos\theta\,\partial_x p + \sin\theta\,\partial_y p + \frac{\sigma^2}{2}\partial^2_{\theta\theta} p = 0 \qquad (2.12)$$

known as "Mumford equation". It is an ultraparabolic equation of Hörmander's type, quite different from those written in our previous examples, and exhibits interesting properties. As to the fundamental solution of this equation, Mumford writes (see [18, p. 497]):

> I have looked for an explicit formula for p but in vain. Still, on the basis of the results of §32.2, I would conjecture that a formula exists, in terms of elliptic functions of some kind.

Let us check Hörmander's condition for (2.12):

$$X_1 = \frac{\sigma}{\sqrt{2}}\partial_\theta; \quad X_0 = (\cos\theta)\,\partial_x + (\sin\theta)\,\partial_y + \partial_t;$$

$$[X_1, X_0] = \frac{\sigma}{\sqrt{2}}\left(-(\sin\theta)\,\partial_x + (\cos\theta)\,\partial_y\right)$$

$$[X_1, [X_1, X_0]] = \frac{\sigma^2}{\sqrt{2}}\left((-\cos\theta)\,\partial_x - (\sin\theta)\,\partial_y\right)$$

and the 4 vector fields $X_1, X_0, [X_1, X_0], [X_1, [X_1, X_0]]$ span $\mathbb{R}^4$ at any point.

Image processing, "the process of random curvature" (See August-Zucker [1]). In the attempt of proposing a new better method for the enhancement of contours of noisy images, August and Zucker have formulated another stochastic model, which we could call "process of random curvature". The setting is similar to that studied by Mumford (see the previous example) but now it is the curvature $\kappa(s)$ to be proportional to a Wiener process. The stochastic system is:

$$\begin{cases} dx = \cos\theta\,ds \\ dy = \sin\theta\,ds \\ d\theta = \kappa\,ds \\ d\kappa = \sigma\,dW \end{cases}$$

and the corresponding backward Kolmogorov equation is

$$\partial_t p + \cos\theta\,\partial_x p + \sin\theta\,\partial_y p + \kappa\,\partial_\theta p + \frac{\sigma^2}{2}\partial^2_{\kappa\kappa} p = 0.$$

We can still check that the vector fields

$$X_1 = \frac{\sigma}{\sqrt{2}}\partial_\kappa; \; X_0 = \text{sum of first order terms}$$

satisfy Hörmander's condition (this time, at step 4).

Other examples. We just quote some other examples from different areas of applied sciences, and corresponding reference:
Diffusion across potential barriers (see [26, Chap. 8]);
Statistical properties of laser light (see [25, Chap. 12]);
Filtering theory (see [26, Chap. 9]);
Dispersive groundwater flow and pollution (see [11, Chap. 6]);
Extinction in systems of interacting biological populations (see [11, Chap. 7]);
Dynamics of polymers (see [12, Chap. 9, Sect. 9.1]);
Astronomy, distribution of "clusters" in space (see [4, 6]).

Let us end with a couple of examples of purely mathematical interest:

Nonhypoelliptic Kolmogorov equation. Consider the system of stochastic O.D.E.:

$$\begin{cases} dx = (x+z)\,dt \\ dy = (y+z)\,dt \\ dz = dw \end{cases}$$

in the variable (x, y, z), where dw is a white noise. The corresponding backward Kolmogorov equation is

$$\partial_t p + (x+z)\,\partial_x p + (y+z)\,\partial_y p + \frac{1}{2}\partial_{zz}^2 p = 0$$

which is ultraparabolic and *does not* satisfy Hörmander's condition. Namely, if

$$X_1 = \partial_t + (x+z)\,\partial_x + (y+z)\,\partial_y; \; X_2 = \frac{1}{\sqrt{2}}\partial_z,$$

$$[X_2, X_1] = X_3 \equiv \frac{1}{\sqrt{2}}\left(\partial_x + \partial_y\right); \quad [X_3, X_1] = X_3$$

so the dimension of the Lie algebra generated by X_1, X_2 is constantly equal to 3, and the equation is not hypoelliptic. (Recall that in the constant rank case, Hörmander's condition is *necessary* for hypoellipticity, see Sect. 1.3.4). Under this respect, this equation is even worse than (2.11).

A stochastic system giving an ultraparabolic equation which satisfies Hörmander's condition at arbitrarily high order. The hypoelliptic, strongly ultraparabolic equation

$$\partial_t p + x_2\partial_{x_1} p + x_3\partial_{x_2} p + \ldots + x_n\partial_{x_{n-1}} p + \partial_{x_n x_n}^2 p = 0$$

is the backward Kolmogorov equation of the (trivial) stochastic differential system

$$\begin{cases} dx_1 = x_2 dt \\ dx_2 = x_3 dt \\ \dots \\ dx_{n-1} = x_n dt \\ dx_n = dw \end{cases}$$

Its solution can be written just taking n times the antiderivative of the white noise dw. The Kolmogorov equation is of type $X_1^2 + X_0$ with

$$X_1 = \partial_{x_n}; \, X_0 = \partial_t + x_2 \partial_{x_1} + x_3 \partial_{x_2} + \dots + x_n \partial_{x_{n-1}}$$

satisfying Hörmander's condition in $\mathbb{R}^{n+1}$ at step n.

2.2 Second Motivation: PDEs Arising in the Theory of Several Complex Variables

2.2.1 Background on the Cauchy-Riemann Complex

Here we will describe very informally some background of the theory of several complex variables which leads to the study of some partial differential operators which turn out to be hypoelliptic. A good reference for this material is the book by Chen-Shaw [7]. Other references will be quoted when appropriate.

Let D be an open subset of $\mathbb{C}^n$. We say that a function $f : D \to \mathbb{C}$ is *holomorphic* if it is holomorphic in each variable z_j when the other variables are fixed. This is equivalent to asking

$$\partial_{\bar{z}_j} f = 0 \text{ for } j = 1, 2, \dots, n$$

where

$$\partial_{\bar{z}_j} = \frac{1}{2} \left(\partial_{x_j} + i \partial_{y_j} \right), \quad z_j = x_j + i y_j$$

is the so-called Cauchy-Riemann operator. One of the important differences between holomorphic functions in one variable and several variables is the following one. Given any open set $D \subset \mathbb{C}$, we can always find a holomorphic function in D which cannot be extended holomorphically to any $D' \supsetneq D$: it's enough to define $u(z) = \frac{1}{z-z_0}$ with $z_0 \in \partial D$. In contrast with this, a fundamental fact discovered by Hartogs in 1906 is the existence of domains $D \subset \mathbb{C}^n$ (with $n > 1$) such that *any* function which is holomorphic in D can be extended holomorphically to some $D' \supsetneq D$. Namely , Hartogs' theorem (see [7, p. 38]) states that: if $D \subset \mathbb{C}^n$ (with $n > 1$) is a bounded domain and K a compact subset of D such that $D \setminus K$

is connected, then any holomorphic function defined in $D \smallsetminus K$ can be extended holomorphically to D. This poses the problem of characterizing the domains of $\mathbb{C}^n$ which *do not* have this property. These domains, called *domains of holomorphy*, are the natural domains of definitions of holomorphic functions in several variables. More precisely, one of the possible definitions of domain of holomorphy reads as follows[2]:

Definition 23 *A domain $D \subset \mathbb{C}^n$ (with $n > 1$) is called a domain of holomorphy if there exists a holomorphic function f on D which is singular at every boundary point.*

The search of a characterization of domains of holomorphy, at increasing levels of generality (domains in $\mathbb{C}^n$, domains in a complex manifold...) is a major theme in the theory of functions of several complex variables. It turns out that a domain of holomorphy in $\mathbb{C}^n$ necessarily possesses a geometric property called *pseudoconvexity*. This is a key concept, which however will not be defined here.[3] Proving the converse implication, namely that any pseudoconvex domain in $\mathbb{C}^n$ is a domain of holomorphy, is known as *the Levi problem*, and has been positively carried out in 1953/54 by Oka, Bremermann, Norguet.[4] In proving this it is also shown the equivalence with a solvability result on the domain D for the inhomogeneous Cauchy-Riemann equation,

$$\overline{\partial} u = f. \tag{2.13}$$

This equation is a central object in the theory of several complex variables. So, let us go into some details about the meaning of (2.13). It is customary to formulate the equation by means of differentials instead of derivatives, and make use of the language of forms. For a complex valued function $u(z_1, z_2, \ldots, z_n)$, one poses:

$$\overline{\partial} u = \sum_{k=1}^{n} \partial_{\overline{z}_k} u \, d\overline{z}_k,$$

hence in (2.13) the assigned f is a one-form

$$f = \sum_{k=1}^{n} f_k \, d\overline{z}_k$$

which, however, must satisfy the compatibility condition

$$\overline{\partial} f = 0,$$

---

[2] This is not the standard definition, but is shown to be equivalent to it. See [7, p. 52–53].

[3] We suggest the short survey paper [24] as an introduction to this idea. Here we just say that pseudoconvexity can be defined via the *Levi form*, which is a kind of "complex hessian" of the defining function of the domain.

[4] See [7, p. 85] for precise references.

since

$$\overline{\partial} f = \overline{\partial} \left(\overline{\partial} u \right) = \overline{\partial} \left(\sum_{k=1}^{n} \partial_{\overline{z}_k} u \, d\overline{z}_k \right) = \sum_{h,k=1}^{n} \partial^2_{\overline{z}_h \overline{z}_k} u \, d\overline{z}_h \wedge d\overline{z}_k$$

$$= \sum_{h>k=1}^{n} \left(\partial^2_{\overline{z}_h \overline{z}_k} u - \partial^2_{\overline{z}_k \overline{z}_h} u \right) d\overline{z}_h \wedge d\overline{z}_k = 0.$$

Generalizing, one considers the space $C^{\infty}_{(p,q)} (D)$ of forms

$$f = \sum_{I,J} {}' f_{I,J} dz^I \wedge d\overline{z}^J$$

where $I = (j_1, j_2, \ldots, j_p)$, $J = (j_1, j_2, \ldots, j_q)$ are multiindices,

$$dz^I = dz_{j_1} \wedge dz_{j_2} \wedge \ldots \wedge dz_{j_p}; d\overline{z}^J = d\overline{z}_{j_1} \wedge d\overline{z}_{j_2} \wedge \ldots \wedge d\overline{z}_{j_q};$$

the sum $\sum'$ is made over increasing multiindices and the $f_{I,J} \in C^{\infty} (D)$ are defined for arbitrary I, J so that they are antisymmetric.

Then, letting

$$\overline{\partial} f = \sum_{I,J} {}' \sum_{j=1}^{n} \partial_{\overline{z}_j} f_{I,J} \, d\overline{z}_j \wedge dz^I \wedge d\overline{z}^J$$

one sees that

$$\overline{\partial} : C^{\infty}_{(p,q)} (D) \longrightarrow C^{\infty}_{(p,q+1)} (D) \tag{2.14}$$

for any $p = 0, 1, \ldots, n$, and

$$0 \longrightarrow C^{\infty}_{(p,0)} (D) \xrightarrow{\overline{\partial}} C^{\infty}_{(p,1)} (D) \xrightarrow{\overline{\partial}} C^{\infty}_{(p,2)} (D) \xrightarrow{\overline{\partial}} \ldots \xrightarrow{\overline{\partial}} C^{\infty}_{(p,n)} (D) \xrightarrow{\overline{\partial}} 0$$

is a complex, that is $\overline{\partial}^2 = 0$. This is *the Cauchy-Riemann complex*. We also write $\overline{\partial}_{p,q}$ instead of $\overline{\partial}$ to denote the single map (2.14). The relation $\overline{\partial}^2 = 0$ means that $R \left(\overline{\partial}_{p,q} \right) \subset Ker \left(\overline{\partial}_{p,q+1} \right)$. If the equality $R \left(\overline{\partial}_{p,q} \right) = Ker \left(\overline{\partial}_{p,q+1} \right)$ holds for any q then we say that *the sequence is exact*, which also means that for any $f \in C^{\infty}_{(p,q+1)} (D)$ satisfying the compatibility condition $\overline{\partial} f = 0$ there exists one solution $u \in C^{\infty}_{(p,q)} (D)$ to the equation $\overline{\partial} u = f$, and u is determined up to holomorphic forms $v \in C^{\infty}_{(p,q)} (D)$ (that is, forms satisfying $\overline{\partial} v = 0$). This happens if D is pseudoconvex (see [7, Thm. 4.5.2 p. 84]).

2.2.2 The $\bar{\partial}$-Neumann Problem

Starting from the Cauchy-Riemann complex it is possible to define a Laplacian, as follows. Let D be a smooth domain in $\mathbb{C}^n$. We want to rewrite the Cauchy-Riemann complex for forms with coefficients in $L^2(D)$, instead of $C^\infty(D)$. First of all, we consider the spaces, denoted by $L^2_{(p,q)}(D)$ (instead of $C^\infty_{(p,q)}(D)$ as before), of forms with $L^2(D)$ coefficients. We can endow $L^2_{(p,q)}(D)$ with the inner product

$$\langle f, g \rangle = \sum_{I,J}{}' \langle f_{I,J}, g_{I,J} \rangle,$$

with the obvious meaning of symbols, so that $L^2_{(p,q)}(D)$ becomes a Hilbert spaces. This allows to introduce a Hilbert space machinery: one takes the (weak) L^2 closure of the unbounded differential operator $\bar{\partial}$, still denoted by $\bar{\partial}$; then for each of the operators

$$\bar{\partial}_{p,q-1} : L^2_{(p,q-1)}(D) \longrightarrow L^2_{(p,q)}(D)$$

one defines the adjoint

$$\bar{\partial}^*_{p,q} : L^2_{(p,q)}(D) \longrightarrow L^2_{(p,q-1)}(D)$$

and, finally, the *Laplacian*

$$\Box_{(p,q)} : L^2_{(p,q)}(D) \longrightarrow L^2_{(p,q)}(D)$$
$$\Box_{(p,q)} = \bar{\partial}_{p,q-1}\bar{\partial}^*_{p,q} + \bar{\partial}^*_{p,q+1}\bar{\partial}_{p,q}. \tag{2.15}$$

The subtle point in the definition of this operator is the identification of its domain. Actually, requiring that a form belongs to the domain of the adjoint operator $\bar{\partial}^*_{p,q}$ implicitly imposes some boundary conditions, of Neumann type, on the form. We will not specify precisely these conditions here (see for instance [7, p. 64–65]). Hence the equation

$$\Box_{(p,q)}U = f \tag{2.16}$$

is actually equivalent to a system of equations *and* boundary conditions (of Neumann type). Therefore (2.16) is called the $\bar{\partial}$-Neumann problem, or the Neumann problem for the Cauchy-Riemann complex. It can be proved (see [7, Prop. 4.2.4, p. 66]) that if $f \in C^2_{(p,q)}(D)$ and $f \in Dom\left(\Box_{(p,q)}\right)$, then

$$\Box_{(p,q)}f = -\frac{1}{4}\sum_{I,J}{}' \Delta f_{I,J}dz^I \wedge d\bar{z}^J$$

where Δ is the classical Laplacian. This justifies the name of Laplacian given also to the operator $\Box$. The $\overline{\partial}$-Neumann problem has been introduced by Garabedian and Spencer in 1952 (see [7, p. 85] for exact references) to study the Cauchy-Riemann equation. Namely, if U solves the $\overline{\partial}$-Neumann problem (2.16) (in particular, we are also assuming that U satisfies the suitable boundary conditions) then $u = \overline{\partial}^*_{p,q} U$ solves the inhomogeneous Cauchy-Riemann equation $\overline{\partial}u = f$. In this sense the $\overline{\partial}$-Neumann problem can be seen as an auxiliary problem to solve the inhomogeneous Cauchy-Riemann equation.[5]

A relevant feature of the $\overline{\partial}$-Neumann problem is that, even though $\Box$ acts on smooth forms like an elliptic operator, the boundary conditions are not elliptic (they do not satisfy Lopatinski condition[6]), which makes the operator $\Box$ itself nonelliptic. To put it into another way, if we define the bilinear form on $L^2_{(p,q)}(D)$ given by

$$Q(\phi, \psi) = \left(\overline{\partial}\phi, \overline{\partial}\psi\right) + \left(\overline{\partial}^*\phi, \overline{\partial}^*\psi\right) + (\phi, \psi),$$

then Q is not coercive on $H^{1,2}(D)$, without some assumption on D. Nevertheless, in 1963 Kohn [13] proved that, if the domain D is *strongly pseudoconvex*,[7] coercivity holds, and the operator $\Box$ satisfies *subelliptic estimates* like (1.6), with $\varepsilon = 1/2$:

$$\|\phi\|^2_{1/2} \le c\left\{|(\Box\phi, \phi)| + \|\phi\|^2_0\right\}$$

Actually, this was the first appearance of subelliptic estimates in the literature.

2.2.3 The Tangential Cauchy-Riemann Complex and the Kohn Laplacian $\Box_b$

Starting from the Cauchy-Riemann complex it is now possible to define the *tangential Cauchy-Riemann complex*, which has been introduced by Kohn-Rossi in 1965. For its clarity and brevity, we quote the next paragraph from Stein's book [27, pp. 592–593].

> The importance of this is made evident by the remarkable principle (of Bochner, Lewy, and others) that a function on the boundary of a suitable domain D is the restriction of some holomorphic function on D exactly when it satisfies the tangential Cauchy-Riemann equations. These considerations give rise to the $\overline{\partial}_b$-complex, which is the boundary analogue of the $\overline{\partial}$-complex discussed above. The idea behind the definition of $\overline{\partial}_b$ is as follows. Suppose

---

[5] See also [27, Chap. 13, Sect. 1] for this point.

[6] See [15, p. 98] for the meaning of this condition.

[7] This is a stronger notion than pseudoconvexity. Let us state a theorem which characterizes it (See [7, Corollary 3.4.5]):

Let D be a bounded pseudoconvex domain with C^2 boundary in $\mathbb{C}^n$, $n \ge 2$. Then D is strongly pseudoconvex if and only if D is locally biholomorphically equivalent to a strictly convex domain near every boundary point.

first that f is a smooth function defined on the boundary of D (i.e., f is a 0-form on ∂D). To define $\overline{\partial}_b f$, we first extend f to F, a smooth function on all of $\overline{D}$, and form $\overline{\partial}_b F$. We then restrict $\overline{\partial}_b F$ to ∂D, and focus on the part of $\overline{\partial}_b F_{/\partial D}$ that is independent of the particular extension F of f. Clearly, this is given by a combination of components that involve only tangential differentiation on the boundary. (...) Boundary q-forms can be defined similarly when $q > 1$, as well as the operator $\overline{\partial}_b$, which maps q-forms to $(q+1)$-forms. In the presence of a suitable inner product, we can also define the formal adjoint $\overline{\partial}_b^*$ and pass to the corresponding boundary Laplacian[8]

$$\Box_b = \overline{\partial}_b \overline{\partial}_b^* + \overline{\partial}_b^* \overline{\partial}_b.$$

Inverting $\Box_b$ is closely related to solving the $\overline{\partial}$-Neumann problem, but is not identical to it. Note that, for $\Box_b$, there are no "boundary conditions"; however the nonellipticity of the $\overline{\partial}$-Neumann problem is reflected here in that the second order operator $\Box_b$ is not elliptic (...): there is always a missing direction.

The operator $\Box_b$ is usually called *the Kohn Laplacian*. Again, $1/2$-subelliptic estimates have been proved for $\Box_b$ on ∂D when D is a strongly pseudoconvex domain. This implies that *the Kohn Laplacian $\Box_b$ on the boundary of a strongly pseudoconvex domain is hypoelliptic.*

2.2.4 The Kohn Laplacian on the Heisenberg Group

Let us now specialize the previous construction of $\Box_b$ to an important particular case. We consider the *generalized upper half-plane*

$$\mathcal{U}^n = \left\{ \zeta \in \mathbb{C}^{n+1} : \sum_{j=1}^n |\zeta_j|^2 < \operatorname{Im} \zeta_0 \right\},$$

which is biholomorphically equivalent[9] to the unit ball in $\mathbb{C}^{n+1}$. The set $\mathcal{U}^n$ is a model example of strongly pseudoconvex domain in $\mathbb{C}^{n+1}$.

Let us consider the real hypersurface $\partial \mathcal{U}^n \subset \mathbb{C}^{n+1}$

---

[8] Here we skip the indices (p, q) of the operators $\overline{\partial}_b$ and $\overline{\partial}_b^*$, to keep more readable the expression. The right indices are the same as in (2.15).

[9] via the following mapping:

$$w_{n+1} = \frac{i - z_{n+1}}{i + z_{n+1}}; \, w_k = \frac{2i z_k}{i + z_{n+1}}, k = 1, 2, \ldots, n.$$

$$z_{n+1} = i \left(\frac{1 - w_{n+1}}{1 + w_{n+1}} \right); z_k = \frac{w_k}{1 + w_{n+1}}, k = 1, 2, \ldots, n.$$

Note that for $n = 1$ this map realizes the biholomorphic equivalence between the halfplane Im $z > 0$ and the circle $|w| < 1$ in the complex plane.

$$\partial \mathcal{U}^n = \left\{ \zeta \in \mathbb{C}^{n+1} : \sum_{j=1}^{n} |\zeta_j|^2 = \operatorname{Im} \zeta_0 \right\}.$$

We want to define $\Box_b$ on $\partial \mathcal{U}^n$. We can identify $\partial \mathcal{U}^n$ with $\mathbb{R}^{2n+1}$ this way: if we set

$$z_j = x_j + iy_j; z = (z_1, z_2, \ldots, z_n)$$

then we identify the point

$$(x_1, x_2, \ldots, x_n, y_1, y_2, \ldots, y_n, t) \in \mathbb{R}^{2n+1}$$

with the point

$$\left(z, t + i |z|^2\right) \in \partial \mathcal{U}^n.$$

Second, we put in $\mathbb{R}^{2n+1}$ the following group law:

$$(z, t) \circ (z', t') = \left(z + z', t + t' + 2 \operatorname{Im} \sum_{j=1}^{n} z_j \overline{z}'_j \right).$$

The set $\mathbb{R}^{2n+1}$ with this group law is called the *Heisenberg group* $\mathbb{H}^n$. The identification of $\partial \mathcal{U}^n$ with $\mathbb{H}^n$ has a geometric meaning, because $\mathbb{H}^n$ allows to define a family of translations in $\mathbb{C}^{n+1}$ which preserve both $\mathcal{U}^n$ and its boundary. Namely, to each $(\zeta, t) \in \mathbb{H}^n$ we associate the following holomorphic affine self-mapping of $\mathcal{U}^n$:

$$T_{(\zeta, t)} : \left(z', z_{n+1}\right) \mapsto \left(z' + \zeta, z_{n+1} + t + 2iz' \cdot \overline{\zeta} + i |\zeta|^2\right).$$

The vector fields:

$$X_j = \partial_{x_j} + 2y_j \partial_t; Y_j = \partial_{y_j} - 2x_j \partial_t; T = \partial_t$$

form a basis for the Lie algebra of $\partial \mathcal{U}^n$. The forms $d\overline{z}_1, d\overline{z}_2, \ldots, d\overline{z}_n$ are a left-invariant basis for the $(0, 1)$-forms on $\partial \mathcal{U}^n$. The operator $\overline{\partial}_b$ is a left-invariant operator on $\partial \mathcal{U}^n$ which can be computed explicitly. If we set

$$Z_j = \frac{1}{2} \left(X_j - iY_j \right) = \partial_{z_j} + i\overline{z}_j \partial_t$$

then we have (with multiindex notation)

$$\overline{\partial}_b \left(\sum_J \phi_J d\overline{z}^J \right) = \sum_J \sum_{k=1}^{n} \left(\overline{Z}_k \phi_J \right) d\overline{z}_k \wedge d\overline{z}^J.$$

Imposing on $\partial \mathcal{U}^n$ a left-invariant metric which makes $Z_1, Z_2, \ldots, Z_n, \overline{Z}_1, \overline{Z}_2, \ldots,$ $\overline{Z}_n$ orthonormal, one can make an explicit computation of the Kohn Laplacian on $(0, q)$-forms:

Theorem 24 *For any $q = 0, 1, 2, \ldots, n$, we have*

$$\Box_b \left(\sum_{|J|=q} \phi_J d\overline{z}^J \right) = \sum_{|J|=q} (\mathcal{L}_\alpha \phi_J) \, d\overline{z}^J \ \text{ with } \alpha = n - 2q$$

where

$$\mathcal{L}_\alpha = -\frac{1}{2} \sum_{k=1}^n \left(Z_k \overline{Z}_k + \overline{Z}_k Z_k \right) + i\alpha T = -\frac{1}{2} \sum_{k=1}^n \left(X_k^2 + Y_k^2 \right) + i\alpha T.$$

The study of $\Box_b$ on q-forms is therefore reduced to the study of the scalar operators $\mathcal{L}_\alpha$ for $\alpha = n, n - 2, n - 4, \ldots, -n$.

Remark 25 *For $\alpha = 0$, $\mathcal{L}_\alpha$ is a sum of squares of Hörmander's vector fields, therefore both hypoelliptic and locally solvable. It is usually called the* sublaplacian *on the Heisenberg group. This is by far the most studied among Hörmander's operators, and we will say more about it in the following.*

Even though the $\mathcal{L}_\alpha$'s involved in $\Box_b$ are only those for $\alpha = n, n - 2, n - 4, \ldots, -n$, these operators have been studied more generally for any $\alpha \in \mathbb{C}$, encountering interesting phenomena:

Theorem 26 *There exists a sequence of forbidden values*

$$\alpha = \pm n, \pm (n + 2), \pm (n + 4), \ldots$$

such that for any admissible (=not forbidden) α the operator $\mathcal{L}_\alpha$ is both locally solvable and hypoelliptic.

Remark 27 *The comparison of different operators $\mathcal{L}_\alpha$ having admissible or forbidden values exhibits interesting examples of operators which have a similar structure but are or are not hypoelliptic and solvable. Note that whenever α is not purely imaginary, $\mathcal{L}_\alpha$ is not a Hörmander's operator because has not real coefficients. Nevertheless, it is hypoelliptic and solvable for almost all values of α.*

Corollary 28 *When $0 < q < n$ the operator $\Box_b$ is both locally solvable and hypoelliptic.*

Remark 29 *For $n = 2$ and $q = 0$ the Kohn Laplacian acts on 0-forms, that is functions, and one finds*

$$\Box_b = -Z\overline{Z}$$

where $\overline{Z}$ is the nonsolvable Lewy operator. In this case $\square_b$ is not solvable. Namely:

$$\overline{Z} = \frac{1}{2}\,(X + iY) = \frac{1}{2}\left[(\partial_x + 2y\partial_t) + i\,(\partial_y - 2x\partial_t)\right]$$
$$= \frac{1}{2}\left[(\partial_x + i\partial_y) - 2i\,(x+iy)\,\partial_t\right],$$

which is exactly Lewy's operator. (See §1.2). Actually, Folland-Stein note in [10, p. 436] that Lewy was led to his example by considering the boundary values of holomorphic functions on $\mathcal{U}^1$.

Summarizing: We have seen how in the study of problems for functions of several complex variables one is naturally lead to study the Kohn Laplacian which, if the domain is strongly pseudoconvex, is a nonelliptic but subelliptic operator. Moreover, if the domain is the generalized upper halfplane $\mathcal{U}^n$, the Kohn Laplacian on $\partial\mathcal{U}^n$ (which can be identified with the Heisenberg group $\mathbb{H}^n$) can be analyzed in terms of operators which are sum of squares of Hörmander's vector fields (sublaplacians), plus a complex drift.

A final important remark is then the following. Any strongly pseudoconvex domain in $\mathbb{C}^n$ can be well approximated (near a boundary point) by $\mathcal{U}^n$. Therefore the Heisenberg group gives the simplest model of the boundary of a strongly pseudoconvex domain, but also a concrete approximation of it. As a result, the exact formulas that hold in that special case can be extended as approximate identities in more general situations. This is a reason why the study of the sublaplacian on the Heisenberg group is so important. Moreover, as we will see in Chap. 3, the subsequent progress in the theory of Hörmander's operators has greatly extended this analogy, giving importance to the study of more general sublaplacians on suitable groups as models and approximations for even more general operators.

References

1. J. August, S.W. Zucker: Sketches with Curvature: the curve indicator random field and Markov processes. IEEE Trans. Pattern Analy. Mach. Intell. **25**(4), 387–400 (2003)
2. Barchielli, A.: Modelli stocastici ed equazioni di Kolmogorov-Fokker-Planck (Italian). Notes for the Summer school on: Campi vettoriali di Hörmander, equazioni differenziali ipoellittiche e applicazioni. 12/16 luglio 2004. http://www.mate.polimi.it/scuolaestiva/bibliografia/barchielli_EDS_FP.pdf
3. Barucci, E., Polidoro, S., Vespri, V.: Some results on partial differential equations and Asian options. Math. Models Methods Appl. Sci. **11**(3), 475–497 (2001)
4. Benacquista, M.: Relativistic Binaries in Globular Clusters. Living reviews in relativity (2002). http://www.livingreviews.org/
5. Brown, R.: A brief account of microscopical observations made in the months of June, July, August, 1827, on the particles contained in the pollen of plants. Phil. Mag. (4), 1828, p. 161–173. Ann. Phys. Chem. **14**, 294 (1828)
6. Chandrasekhar, S.: Rev. Mod. Phys. **15**, 1 (1943). This paper is also contained in [28]

7. Chen, S.C., Shaw, M.C.: Partial differential equations in several complex variables. AMS/IP studies in advanced mathematics, vol. 19. American Mathematical Society, Providence, RI. International Press, Boston, (2001)
8. Einstein, A.: Ann. Physik **17**, 549 and 19, 371 (1906). These and other papers by Einstein are collected in: Einstein. A (ed.), 2nd edn in 1956. Investigations on the Theory of Brownian Movement. Dover, New York (1905)
9. Fokker, A.D.: Ann. Physik **43**, 810 (1914)
10. Folland, G.B., Stein, E.M.: Estimates for the $\bar{\partial}_b$ complex and analysis on the Heisenberg group. Comm. Pure Appl. Math. **27**, 429–522 (1974)
11. Grasman, J., van Herwaarden, O.A.: Asymptotic Methods for the Fokker-Planck Equation and the Exit Problem in Applications. Springer, Berlin (1999)
12. Honerkamp, J.: Stochastic Dynamical Systems. VCH, Weinheim (1994)
13. Kohn, J.J.: Regularity at the boundary of the $\bar{\partial}$-Neumann problem. Proc. Nat. Acad. Sci. USA **49**, 206–213 (1963)
14. Kolmogorov, A.N.: Math. Ann. **104**, 415–458 (1931)
15. Krantz, S.G.: Partial differential equations and complex analysis. Studies in Advanced Mathematics. CRC press, Boca Rato (1992)
16. Langevin, P.: C. R. **146**, 530 (1908)
17. Malliavin, P.: Stochastic calculus of variations and hypoelliptic operators. In: Proceedings International Symposium on Stochastic Differential Equations, pp. 195–263. Wiley, Kyoto 1976 (1978)
18. Mumford, D.: Elastica and computer vision. Algebraic geometry and its applications, pp. 491–506. In: Springer, West Lafayette 1990 (1994)
19. Nualart, D.: The Malliavin Calculus and Related Topics. Springer-Verlag, Berlin (1995)
20. Ornstein, L.S., Uhlenbeck, G.E.: Phys. Rev. **36**, 823 (1930). This paper is also contained in [28]
21. Pascucci, A.: Calcolo stocastico per la finanza. Springer 2008
22. Pascucci, A.: Kolmogorov equations in physics and in finance. Elliptic and parabolic problems, vol. 63, pp. 353–364. Progr. Nonlinear Differential Equations Application. Birkhäuser, Basel (2005)
23. Planck, M.: Sitzber. Preuss. Akad. Wiss. **324**, (1917)
24. Range, M.R.: What is... a pseudoconvex domain? Notices of the A.M.S. February 2012, 59(02), pp. 301–303 (2012). http://www.ams.org/notices/201202/rtx120200301p.pdf
25. Risken, H.: The Fokker-Planck equation. Methods of solution and applications. Springer Series in Synergetics, vol. 18, 2nd edn. Springer-Verlag, Berlin (1989)
26. Schuss, Z.: Theory and Applications of Stochastic Differential Equations. Wiley, New York (1980)
27. Stein, E.M.: Harmonic analysis: real-variable methods, orthogonality, and oscillatory integrals. With the assistance of Timothy S. Murphy. Princeton mathematical series, vol. 43. Monographs in Harmonic Analysis, III. Princeton University Press, Princeton (1993)
28. Wax, N. (ed.): Selected Papers on Noise and Stochastic Processes. Dover, New York (1954)
29. Wiener, N.: Differential space. J. Math. Physics **2**, 131–174 (1923)

Chapter 3
A Priori Estimates in Sobolev Spaces for Hörmander's Operators

3.1 What are the "Natural" a Priori-Estimates to be Proved for Hörmander's Operators?

We have discussed in Sect. 1.3.5 Kohn's approach to the proof of Hörmander's theorem. First, for an operator L of Hörmander's type, one proves the following *subelliptic estimates* in fractional Sobolev spaces of small positive order:

$$\|u\|_{H^{\varepsilon,2}} \leq c \left(\|Lu\|_{L^2} + \|u\|_{L^2} \right)$$

for any $u \in C_0^\infty (\mathbb{R}^n)$, some $\varepsilon > 0$. Then, one gets the analogous higher order estimate

$$\|u\|_{H^{s+\varepsilon,2}} \leq c_{m,s} \left(\|Lu\|_{H^{s,2}} + \|u\|_{H^{-m,2}} \right)$$

for any $s, m > 0$. These imply hypoellipticity. The proof of these estimates involve techniques of pseudodifferential operators, hence the Fourier transform on L^2, Hilbert space techniques, Sobolev spaces of fractional order.

Subelliptic estimates open a natural problem. We are able to bound just a fractional derivative of u in terms of Lu, even though the operator L is highly regularizing. This is somehow unsatisfactory. To put it into another way: we know that $Lu \in C^\infty (\Omega)$ implies $u \in C^\infty (\Omega)$ but if Lu has some partial regularity, for instance $Lu \in H^{k,2}$ for *some* (but not for all) k, or if $Lu \in L^p (\Omega)$ for some $p \neq 2$, subelliptic estimates do not allow us to deduce the natural gain of regularity of u. The point is that we are trying to bound the usual, "Cartesian" derivatives for an operator which is highly anisotropic: we can expect L to control the specific directions given by the vector fields X_i. So perhaps the situation could be better if we tried to bound the derivatives $X_j u$ and $X_i X_j u$. But this requires a completely different approach: the techniques of pseudodifferential operators used to get subelliptic estimates are shaped on *fractional* but *isotropic* derivatives; moreover, they privilege L^2 bounds with respect to L^p bounds for other p's; also, if one tries to get a bound on the derivatives with respect to the vector fields, the Hilbert space technique will offer an

M. Bramanti, *An Invitation to Hypoelliptic Operators and Hörmander's Vector Fields*, SpringerBriefs in Mathematics, DOI: 10.1007/978-3-319-02087-7_3, © The Author(s) 2014

estimate on *first order derivatives*, like in the following "energy estimate" that we can prove for the sublaplacian on $\mathbb{H}^1$:

$$\|Xu\|^2 + \|Yu\|^2 \leq \left| \int Lu \cdot u \right| \leq \|Lu\| \cdot \|u\| \lesssim \|Lu\|^2 + \|u\|^2 .$$

Instead, if one wants to prove an L^p bound on $X_i X_j u$ (and $X_0 u$, if the drift is present) in terms of Lu and u, then one should mimic the techniques used to prove L^p estimates for *strong solutions to nonvariational elliptic equations*. This involves the use of representation formulas by means of fundamental solutions, and the application of singular integral estimates (Calderón-Zygmund theory). We will recall more about these topics later. This program has been actually carried out, for Hörmander's operators, in three famous, outstanding papers of the mid-seventies:

1974, Folland-Stein, Comm. Pure Appl. Math., [12]

1975, Folland, Arkiv für Mat., [9]

1976, Rothschild-Stein, Acta Math., [19]

In these papers a priori estimates of the kind[1]

$$\left\| X_i X_j u \right\|_p + \|X_0 u\|_p \lesssim \|Lu\|_p + \|u\|_p \text{ for } 1 < p < \infty$$

have been proved, at increasing levels of generality. Namely:

Step 1: 1974, Folland-Stein: the Kohn-Laplacian on Heisenberg groups (part I) and on nondegenerate CR manifolds (part II);

Step 2: 1975, Folland: sublaplacians on homogeneous groups;

Step 3: 1976, Rothschild-Stein: general Hörmander's operators.

These papers introduced a number of fundamental ideas which, still now, represent some of the basic tools which are necessary in order to do research in this field.

We want to stress the fact that each further step does not make the previous ones "useless". Namely: the results of Step 3 extend those of Step 2, but are also based on them; the main result of part I of Step 1 is in some sense properly contained in Step 2; however, part II of Step 1 is not contained in Step 2; it consists in an extension of the results of Part I to a more general situation, and that way of reasoning can be considered as the seed of the ideas used in Step 3 to extend Step 2.

The rest of this chapter will be devoted to explaining, in the general lines, some of the main ideas contained in these papers. Although, from the standpoint of the subsequent development of the theory, the most important points to be learned are contained in steps 2 and 3, we will remain for a while on step 1, which is a useful preparation to step 2.

I suggest to the interested reader also the survey paper [11] by Folland, 1977, which contains a good introduction to the same three papers.

---

[1] The exact form of these a priori estimates will be made precise in the following, dealing with each of the three papers.

3.2 The Sublaplacian on the Heisenberg Group

Here we will discuss some ideas contained in the paper by Folland-Stein [12], 1974, with the necessary background. This will also serve as an introduction to the Heisenberg group and to the first ideas about homogeneous groups, which we will present in more depth in the next section.

As we have explained in Sect. 2.2.4, the study of the Kohn Laplacian $\Box_b$ (which is an operators acting on forms) on the boundary $\partial \mathcal{U}^n$ of the generalized upper halfplane $\mathcal{U}^n \subset \mathbb{C}^{n+1}$ can be reduced to the study of the following differential operators in $\mathbb{R}^{2n+1}$ (acting on functions):

$$\mathcal{L}_\alpha = -\frac{1}{2} \sum_{k=1}^n \left(X_k^2 + Y_k^2 \right) + i\alpha T$$

for $\alpha = n, n-2, n-4, ..., -n$, where

$$X_j = \partial_{x_j} + 2y_j \partial_t;$$
$$Y_j = \partial_{y_j} - 2x_j \partial_t;$$
$$T = \partial_t.$$

Henceforth we will concentrate on the scalar operators $\mathcal{L}_\alpha$, which have been studied by Folland-Stein [12] for every $\alpha \in \mathbb{C}$. The operator $\mathcal{L}_0$ is usually called the sublaplacian on the Heisenberg group.

3.2.1 The Classical Laplacian

Let us first give an overview of some facts related to the classical Laplacian, which will have a natural counterpart dealing with the sublaplacian (or $\mathcal{L}_\alpha$) in the Heisenberg group. The Laplacian

$$\Delta = \sum_{k=1}^n \partial^2_{x_k x_k}$$

in $\mathbb{R}^n$ ($n \geq 3$) is invariant for (Euclidean) translations, that is

$$\Delta_x (f(x+y)) = (\Delta f)(x+y) \text{ for any } y \in \mathbb{R}^n$$

because it has constant coefficients. It is also homogeneous of degree 2 with respect to the usual dilations, that is

$$\Delta (f(\lambda x)) = \lambda^2 (\Delta f)(\lambda x).$$

Finally, it is rotation invariant, that is

$$\Delta\left(f\left(Rx\right)\right) = \left(\Delta f\right)\left(Rx\right)$$

whenever R is an orthogonal $n \times n$ matrix.

The operator $-\Delta$ has a positive fundamental solution which is translation invariant, that is

$$\Gamma\left(x, y\right) = \gamma\left(x - y\right).$$

Actually,

$$\gamma\left(x\right) = \frac{c_n}{|x|^{n-2}},$$

in particular γ is rotation invariant and homogeneous of degree $2 - n$: so we see how properties of the differential operator reflect in the properties of its fundamental solution. One can write the representation formula

$$u\left(x\right) = -\int_{\mathbb{R}^n} \gamma\left(x - y\right) \Delta u\left(y\right) dy \text{ for any } u \in C_0^\infty\left(\mathbb{R}^n\right).$$

Hence one can compute

$$u_{x_i}\left(x\right) = -\int_{\mathbb{R}^n} \gamma_{x_i}\left(x - y\right) \Delta u\left(y\right) dy.$$

Taking one more derivative is troublesome because $\gamma_{x_i x_j}$ is no longer locally integrable, being homogeneous of degree $-n$; anyhow, one finds:

$$u_{x_i x_j}\left(x\right) = -P.V.\int_{\mathbb{R}^n} \gamma_{x_i x_j}\left(x - y\right) \Delta u\left(y\right) dy + c_{ij}\Delta u\left(x\right),$$

where c_{ij} are suitable constants and $P.V.$ stands for the principal value of the singular integral, defined as

$$P.V.\int_{\mathbb{R}^n} \gamma_{x_i x_j}\left(x - y\right) f\left(y\right) dy = \lim_{\varepsilon \to 0}\int_{|x-y|>\varepsilon} \gamma_{x_i x_j}\left(x - y\right) f\left(y\right) dy.$$

Letting

$$T_{ij} f = P.V.\left(\gamma_{x_i x_j} * f\right)$$

we can write

$$\|u_{x_i x_j}\|_{L^p} \le c \|\Delta u\|_{L^p} + \|T_{ij}\left(\Delta u\right)\|_{L^p}$$

and the problem of proving a-priori L^p estimates on $u_{x_i x_j}$ is reduced to that of proving the L^p continuity of the singular integral operator T_{ij}. The continuity of T_{ij} for $p \in (1, \infty)$ is well-known by the classical theory of singular integrals, developed by Calderón and Zygmund in a series of papers starting with [3], 1952, until the late 1970s. Actually, the kernel $k = \gamma_{x_i x_j}$ satisfies the following properties:

$$|k(x)| \le \frac{c}{|x|^n} \text{ ("growth condition");} \tag{3.1}$$

$$|\nabla k(x)| \le \frac{c}{|x|^{n+1}} \tag{3.2}$$

$$\int_{R_1 < |x| < R_2} k(x)\, dx = 0 \text{ ("vanishing property")} \tag{3.3}$$

The second property follows from the first since k is smooth outside the origin and homogeneous of degree $-n$, while (3.3) is easily read by the explicit expression of $\gamma_{x_i x_j}$. We also note that (3.2) implies the following, which is the inequality really needed in proofs:

$$|k(x+y) - k(x)| \le c\frac{|y|}{|x|^{n+1}} \text{ for all } x, y \text{ with } |x| \ge 2|y|. \tag{3.4}$$

The three properties (3.1), (3.4), (3.3) are sufficient (although not necessary) to make Calderón-Zygmund's theory work, granting the L^p continuity of T_{ij} for $p \in (1, \infty)$. This is the backbone of L^p theory for the classical Laplacian. Let us now see how the sublaplacian mimics its classical counterpart.

3.2.2 Geometry of the Sublaplacian

Let us recall that the Heisenberg group $\mathbb{H}^n$ is the set $\mathbb{R}^{2n+1} \ni (z, t)$, where

$$z = (z_1, z_2, ..., z_n), z_j = x_j + iy_j,$$

endowed with the following group law[2]:

$$(z, t) \circ (z', t') = \left(z + z', t + t' + 2\,\mathrm{Im} \sum_{j=1}^{n} z_j \bar{z}'_j \right).$$

The vector fields

$$X_k = \partial_{x_k} + 2y_k \partial_t; \ Y_k = \partial_{y_k} - 2x_k \partial_t, k = 1, 2, ..., n$$

are left invariant and satisfy Hörmander's condition in $\mathbb{R}^{2n+1}$, hence the operator

---

[2] The group law of $\mathbb{H}^n$ has already been defined in Sect. 2.2.4.

$$\mathcal{L}_0 = -\frac{1}{2} \sum_{k=1}^{n} \left(X_k^2 + Y_k^2 \right)$$

is hypoelliptic in $\mathbb{R}^{2n+1}$.

Here we will make some explicit computation in $\mathbb{H}^1$, but what we will say holds more generally in $\mathbb{H}^n$.

1. Let us check that the vector fields

$$X = \partial_x + 2y\partial_t; \, Y = \partial_y - 2x\partial_t; \, T = \partial_t$$

in $\mathbb{R}^3 \ni (x, y, t)$ are left invariant with respect to the translation assigned by the group law in $\mathbb{H}^1$, namely

$$\left(x', y', t' \right) \circ (x, y, t) = \left(x + x', y + y', t + t' + 2 \left(xy' - x'y \right) \right).$$

Left invariance means that, for instance,

$$X_\xi \left(f \left(\eta \circ \xi \right) \right) = (Xf) \left(\eta \circ \xi \right).$$

Let us check this condition for the vector field X.

$$X_{(x,y,t)} \left[f \left(\left(x', y', t' \right) \circ (x, y, t) \right) \right]$$
$$= (\partial_x + 2y\partial_t) \left[f \left(x + x', y + y', t + t' + 2 \left(xy' - x'y \right) \right) \right]$$
$$= (\partial_x f) (...) + (\partial_t f) (...) \cdot 2y' + 2y (\partial_t f) (...)$$
$$= (\partial_x f) (...) + 2 \left(y + y' \right) (\partial_t f) (...)$$
$$= (Xf) \left(\left(x', y', t' \right) \circ (x, y, t) \right).$$

Analogously one checks left invariance of Y and T. By composition of left invariant operators, also *the operator $\mathcal{L}_\alpha$ is left invariant.*

2. Next, let us check that the vector field X is homogeneous of degree 1 with respect to the dilations

$$D(\lambda)(x, y, t) = \left(\lambda x, \lambda y, \lambda^2 t \right) \quad \text{for } \lambda > 0.$$

This means that

$$X_{(x,y,t)} \left[f \left(D(\lambda)(x, y, t) \right) \right] = \lambda (Xf) \left(D(\lambda)(x, y, t) \right).$$

Namely,

$$X_{(x,y,t)}\left[f\left(D\left(\lambda\right)\left(x,y,t\right)\right)\right] = \left(\partial_x + 2y\partial_t\right)\left[f\left(\lambda x, \lambda y, \lambda^2 t\right)\right]$$
$$= \lambda\partial_x f\left(\ldots\right) + 2y\lambda^2\partial_t f\left(..\right)$$
$$= \lambda\left[\partial_x f\left(\ldots\right) + 2\left(\lambda y\right)\partial_t f\left(..\right)\right]$$
$$= \lambda\left(Xf\right)\left(D\left(\lambda\right)\left(x,y,t\right)\right).$$

Analogously one checks that Y is homogeneous of degree 1 and T is homogeneous of degree 2 with respect to the same dilations. There follows that $\mathcal{L}_\alpha$ is *homogeneous of degree 2 with respect to these dilations.*

By the way, it is worthwhile to be noted that "translations" and "dilations" satisfy a mutual relation which is analogous to the Euclidean case: as, in the Euclidean case, one has

$$\lambda\left(u + v\right) = \lambda u + \lambda v,$$

so, in the Heisenberg group,

$$D\left(\lambda\right)\left(u \circ v\right) = \left[D\left(\lambda\right) u\right] \circ \left[D\left(\lambda\right) v\right]$$

that is, the "dilations" $D\left(\lambda\right)$ *form a one parameter family of automorphisms of the group* $(\mathbb{H}^n, \circ)$.

3. The behavior of the operator $\mathcal{L}_\alpha$ with respect to rotations is more subtle. First, one can easily conjecture that the "natural" rotations for $\mathcal{L}_\alpha$ are only the Euclidean rotations which fix the t axis. Now, the operator $\mathcal{L}_\alpha$ is actually invariant with respect to these rotations only in dimension $n = 1$. However, in higher dimension, if we restrict the operator $\mathcal{L}_\alpha$ to functions which are radial in z, that is functions $g\left(|x|^2 + |y|^2, t\right)$, then it can be checked that $\mathcal{L}_\alpha$ is rotation invariant.[3] As we will see, $\mathcal{L}_\alpha$ possesses a fundamental solution of the form $g\left(|x|^2 + |y|^2, t\right)$.

3.2.3 Fundamental Solution of the Sublaplacian

We are looking for analogies between the classical Laplacian and the sublaplacian. The fundamental solution of the Laplacian is (modulo translations)

$$\gamma\left(x\right) = \frac{c_n}{|x|^{n-2}}.$$

Optimistically, one can look for a fundamental solution of $\mathcal{L}_\alpha$ (or at least $\mathcal{L}_0$, which is real valued and perhaps easier) of the same kind. We need to understand what is the analog of the Euclidean norm $|x|$ and what is the analog of the dimension n. The second question is easier: if we see the dilations as a change of variables in an integral, we see that in the Euclidean case

---

[3] This computation is not trivial, and here we will skip it. See for instance the book [2, Sect. 3.2].

$$x' = \lambda x \implies dx' = \lambda^n dx \text{ and } n \text{ is the dimension;}$$

in $\mathbb{H}^1$

$$(x', y', t') = D\,(\lambda)\,(x, y, t) = \left(\lambda x, \lambda y, \lambda^2 t\right) \implies dxdydt = \lambda^4 dxdydt$$

and $Q = 4$ is the *homogeneous dimension*. Analogously in $\mathbb{H}^n$ one find

$$D\,(\lambda)\,(x_1, ..., x_n, y_1, ..., y_n, t) = \left(\lambda x_1, ..., \lambda x_n, \lambda y_1, ..., \lambda y_n, \lambda^2 t\right)$$

and $Q = 2n + 2$.

The analog of the Euclidean norm should be a "homogeneous norm", that is a nonnegative function

$$\|(x, y, t)\|$$

such that

$$\|\xi\| = 0 \iff \xi = 0; \tag{3.5}$$

$$\|D\,(\lambda)\,\xi\| = \lambda \|\xi\| \,\forall \lambda > 0; \tag{3.6}$$

$$\|x \circ y\| \le c\,(\|x\| + \|y\|) \tag{3.7}$$

where the last inequality is the analog of the triangle inequality, with the Euclidean translations replaced by the group translations, and a bit of flexibility in allowing a constant $c \ge 1$ at the right-hand side. Moreover, we want $\|\xi\|$ to be a smooth function outside the origin. Our hope is that, analogously to what happens for the classical Laplacian, a fundamental solution for the sublaplacian be given by

$$\gamma\,(x, y, t) = c\,\|(x, y, t)\|^{2-Q}. \tag{3.8}$$

The choice of the homogeneous norm is not unique. One could choose, for instance

$$\|(x, y, t)\| = \sqrt[4]{x^4 + y^4 + t^2};$$

$$\|(x, y, t)\| = \sqrt[4]{\left(x^2 + y^2\right)^2 + t^2};$$

$$\|(x, y, t)\| = \sqrt{\left(x^2 + y^2\right)} + \sqrt[4]{\left(x^2 + y^2\right)^2 + t^2};$$

$$...$$

or analogous versions in $\mathbb{R}^{2n+1}$ with x, y replaced by $|x|$, $|y|$. All of them are metrically equivalent, but to choose the "right" one is crucial, to get (3.8) fulfilled.

In 1973, Folland [8] proved that letting, in $\mathbb{H}^n$

3.2 The Sublaplacian on the Heisenberg Group

$$\|(x, y, t)\| = \sqrt[4]{\left(|x|^2 + |y|^2\right)^2 + t^2}$$

(with $x = (x_1, x_2, ..., x_n)$, $y = (y_1, y_2, ..., y_n)$), the sublaplacian

$$\mathcal{L}_0 = -\frac{1}{2} \sum_{k=1}^{n} \left(X_k^2 + Y_k^2\right)$$

on the Heisenberg group $\mathbb{H}^n$ admits a left invariant, $(2 - Q)$-homogeneous, rotation invariant[4] fundamental solution (3.8) with $c = c_n$. The analogy with the classical Laplacian is really striking.

On the wave of this success, Folland-Stein in [12] looked for a fundamental solution of the operator

$$\mathcal{L}_\alpha = -\frac{1}{2} \sum_{k=1}^{n} \left(X_k^2 + Y_k^2\right) + i\alpha T,$$

with $\alpha \in \mathbb{C}$. It was already known that this operator possesses some forbidden values of α, namely $\pm\alpha = n, n + 2, n + 4, ...$ (see Thm. 2.8 in Sect. 2.2.4), for which it loses hypoellipticity and solvability, so the quest is: to find a fundamental solution γ_α to

$$\mathcal{L}_\alpha \gamma_\alpha = \delta \text{ for } \pm\alpha \neq n, n + 2, n + 4, ...$$

Then, Folland-Stein look for a γ_α of the form:

$$\gamma_\alpha (z, t) = \|(z, t)\|^{-2n} f\left(\frac{t}{|z|^2}\right)$$

(with $z = (x, y)$) and with explicit computations they solve the resulting ODE proving the following

Theorem 30 *Let*

$$\gamma_\alpha = \|(z, t)\|^{-2(n+\alpha)} \left(|z|^2 + it\right)^\alpha = \left(|z|^2 - it\right)^{-(n+\alpha)/2} \left(|z|^2 + it\right)^{-(n-\alpha)/2}.$$

Then

$$\mathcal{L}_\alpha \gamma_\alpha = c_\alpha \delta$$

with

$$c_\alpha = \frac{2^{2-2n} \pi^{n+1}}{\Gamma\left(\frac{n+\alpha}{2}\right) \Gamma\left(\frac{n-\alpha}{2}\right)}.$$

---

[4] The rotations we are considering are just those around the t-axis.

Since the Euler gamma function $\Gamma(s)$ goes to infinity exactly for $s = 0, -1, -2, \ldots$ we see that $c_\alpha = 0$ precisely when

$$\frac{n \pm \alpha}{2} = 0, -1, -2, \ldots$$

that is when $\pm\alpha = n, n + 2, n + 4, \ldots$, which are exactly the "forbidden values". If $c_\alpha \neq 0$, then γ_α / c_α is a fundamental solution for $\mathcal{L}_\alpha$. For $\alpha = 0$ we get Folland's result for the sublaplacian

$$\gamma_0(z, t) = \|(z, t)\|^{-2n}.$$

Again, the fundamental solution found for $\mathcal{L}_\alpha$ is translation invariant, $2 - Q$ homogeneous and rotation invariant (with the same meaning of these properties, as in the case $\alpha = 0$).

3.2.4 What we can do with a Good Fundamental Solution

For any $u \in C_0^\infty(\mathbb{R}^{2n+1})$ and any admissible value α (but we are particularly interested in the case $\alpha = 0$) we can write

$$u(\xi) = \int_{\mathbb{R}^{2n+1}} \gamma_\alpha\left(\eta^{-1} \circ \xi\right) \mathcal{L}_\alpha u(\eta)\, d\eta.$$

Then, exploiting the left invariance of the vector fields X_j (analogous computations can be done for Y_j)

$$X_j u(\xi) = \int_{\mathbb{R}^{2n+1}} (X_j \gamma_\alpha)\left(\eta^{-1} \circ \xi\right) \mathcal{L}_\alpha u(\eta)\, d\eta.$$

As in the Euclidean case, taking a second derivative is troublesome, but one can prove that

$$X_k X_j u(\xi) = P.V. \int_{\mathbb{R}^{2n+1}} (X_k X_j \gamma_\alpha)\left(\eta^{-1} \circ \xi\right) \mathcal{L}_\alpha u(\eta)\, d\eta + c_{kj} \mathcal{L}_\alpha u(\xi) \quad (3.9)$$

where the convolution kernel

$$K(\xi) = \left(X_k X_j \gamma_\alpha\right)(\xi)$$

satisfies properties which are analogous to those of classical Calderón-Zygmund kernels:

Proposition 31 *The function* $K(\xi) = \left(X_k X_j \gamma_\alpha\right)(\xi)$ *satisfies the following:*

$$|K(\xi)| \leq \frac{c}{\|\xi\|^{Q}}; \tag{3.10}$$

$$|K(\xi \circ \eta) - K(\xi)| + |K(\eta \circ \xi) - K(\xi)| \leq c\frac{\|\eta\|}{\|\xi\|^{Q+1}} \text{ for } \|\xi\| \geq 2\|\eta\|; \tag{3.11}$$

$$\int_{R_1 < \|\xi\| < R_2} K(\xi) \, d\xi = 0 \text{ for any } 0 < R_1 < R_2 < \infty. \tag{3.12}$$

Proof The first property is quite obvious, but it is worthwhile to point out the general reasoning which is involved. One can check that whenever Z is a differential operator of homogeneous degree α, that is

$$Z[f(D(\lambda)\xi)] = \lambda^{\alpha}(Zf)(D(\lambda)\xi)$$

and f is a $D(\lambda)$-homogeneous function of degree β, then Zf is a homogeneous function of degree $\beta - \alpha$. Namely, in this case we have

$$\lambda^{\beta}Zf(\xi) = Z[\lambda^{\beta}f(\xi)] = Z[f(D(\lambda)\xi)] = \lambda^{\alpha}(Zf)(D(\lambda)\xi), \text{ and}$$
$$(Zf)(D(\lambda)\xi) = \lambda^{\beta-\alpha}Zf(\xi).$$

Hence $K = (X_k X_j \gamma_\alpha)$ is $-Q$-homogeneous; its smoothness outside the origin then implies (3.10).

The second property still follows by a homogeneity argument.[5] Namely, if we replace ξ, η with $D(\lambda)\xi, D(\lambda)\eta$, by $-Q$-homogeneity of K we get the same inequality, with both sides multiplied by λ^{-Q}; hence it is enough to prove (3.11) for $\|\xi\| = 1$, $\|\eta\| < 1/2$. Under these assumptions $\xi \circ \eta$ and $\eta \circ \xi$ are bounded away from zero, hence

$$|K(\xi \circ \eta) - K(\xi)| + |K(\eta \circ \xi) - K(\xi)| \leq c|\eta|$$

where $|\eta|$ is the Euclidean norm. However, from the very definition of $\|\eta\|$ we see that $|\eta| \leq c\|\eta\|$ for $\|\eta\| \leq 1$. Hence for these ξ and η we have

$$|K(\xi \circ \eta) - K(\xi)| + |K(\eta \circ \xi) - K(\xi)| \leq c\|\eta\| = c\frac{\|\eta\|}{\|\xi\|^{Q+1}}$$

and we are done.

As to the vanishing property (3.12), it is also interesting the abstract way it can be proved. The function $(X_j \gamma_\alpha)$ is a locally integrable $(1 - Q)$-homogeneous function (this local integrability will be proved here below), hence can also be seen as a $(1 - Q)$-homogeneous distribution; hence its derivative $(X_k X_j \gamma_\alpha)$ is a $(-Q)$-

---

[5] Here we are repeating an argument which is actually contained in the subsequent paper [9], not in [12].

homogeneous distribution. Now, Folland-Stein prove (see [12, Prop. 8.2 and 8.5]) that any $(-Q)$-homogeneous distribution which outside the origin coincides with a smooth function satisfies a vanishing property like (3.12). Note that we could not say the same for a $(-Q)$-homogeneous *function*. In this case, being a *distribution* is better than being a *function*! ∎

By comparison, it is instructive to see how a proof of the vanishing property can be carried out, in the classical Calderón-Zygmund case, using exactly the fact that the kernel K is a derivative of a $(1-n)$-homogeneous function. Let

$$k(x) = \partial_{x_i} f(x) \text{ with } f(\lambda x) = \lambda^{1-n} f(x).$$

Then

$$\int_{R_1 < |x| < R_2} \partial_{x_i} f(x)\, dx = \int_{|x|=R_2} f(x)\, n_i d\sigma(x) - \int_{|x|=R_1} f(x)\, n_i d\sigma(x)$$

with

$$\int_{|x|=R} f(x)\, n_i d\sigma(x) = \int_{|x|=R} f(x)\, \frac{x_i}{R} d\sigma(x) = [x = Ru]$$

$$= \int_{|u|=1} f(Ru)\, u_i R^{n-1} d\sigma(u)$$

$$= \int_{|u|=1} R^{1-n} f(u)\, u_i R^{n-1} d\sigma(u)$$

$$= \int_{|u|=1} f(u)\, u_i d\sigma(u) = c_i$$

hence

$$\int_{R_1 < |x| < R_2} \partial_{x_i} f(x)\, dx = c_i - c_i = 0.$$

Once one knows that

$$\int_{R_1 < |x| < R_2} K(x)\, dx = 0 \ \forall R_1 < R_2,$$

one can also derive that

$$\int_{|x|=R} K(x)\, d\sigma(x) = \lim_{h \to 0} \frac{1}{2h} \int_{R-h < |x| < R+h} K(x)\, dx = 0.$$

It is also instructive to see how the homogeneous dimension Q plays the role of the Euclidean dimension n in deciding the integrability of a function possessing a singularity[6]:

Proposition 32 *For any $\alpha > 0$ there exists $c_\alpha > 0$ such that for any $R > 0$,*

$$\int_{\|u\| \leq R} \frac{du}{\|u\|^{Q-\alpha}} \leq c_\alpha R^\alpha.$$

In particular, any $(\alpha - Q)$-homogeneous function is locally integrable.

Proof Let us write:

$$\int_{\|u\| \leq R} \frac{du}{\|u\|^{Q-\alpha}} = \sum_{k=0}^\infty \int_{\frac{R}{2^{k+1}} < \|u\| \leq \frac{R}{2^k}} \frac{du}{\|u\|^{Q-\alpha}} \leq \sum_{k=0}^\infty \left(\frac{R}{2^{k+1}}\right)^{-Q+\alpha} \int_{\|u\| \leq \frac{R}{2^k}} du$$

$$= \sum_{k=0}^\infty \left(\frac{R}{2^{k+1}}\right)^{-Q+\alpha} c \left(\frac{R}{2^k}\right)^Q = R^\alpha c \sum_{k=0}^\infty \left(\frac{1}{2^{\alpha(k+1)}}\right) = c_\alpha R^\alpha.$$

∎

A slightly different computation also shows that for any $\alpha \geq 0$, $R > 0$

$$\int_{\|u\| \leq R} \frac{du}{\|u\|^{Q-\alpha}} \geq c R^\alpha \sum_{k=0}^\infty \frac{1}{2^{k\alpha}}$$

which also shows the nonintegrability of $\|u\|^{-Q}$.

Let us summarize. We have reduced the proof of a-priori L^p estimates on $X_i X_j u$ to the proof of L^p continuity of a singular integral operator with kernel $X_i X_j \gamma_\alpha$; moreover, analyzing this kernel, we have checked that it satisfies the "standard estimates" (3.10), (3.11) and the vanishing property (3.12), which are perfectly analogous to the properties satisfied by a classical Calderón-Zygmund kernel, with Euclidean translations and dilations replaced by the translations and dilations in the Heisenberg group, that is the homogeneous group which makes the operator $\mathcal{L}_\alpha$ left invariant and 2-homogeneous. We are in a good position to hope that L^p estimates can actually be proved. But theorems are not proved just "by analogy", and Calderón-Zygmund's theory is not an easy one: in order to extend its results to an analogous more general situation, some work must be done. This is actually another story, which we are going to retell very briefly. Before doing so, let us note that the reasoning used to prove that the kernel $X_i X_j \gamma_\alpha$ satisfies suitable properties is actually very general,

---

[6] It is possible to prove a more precise result about the integration of radial functions in homogeneous groups (see [9, Prop. 1.5]), which allows to prove an equality, and not just an inequality. I prefer to present this argument because it is simpler, can be extended to more general situations, and is enough for our purposes.

does not depend much on possessing an explicit form of γ_α; this will be crucial for the subsequent development of the theory in [9].

3.2.5 Singular Integrals in Spaces of Homogeneous Type

To understand the idea behind the extensions of the classical theory of singular integrals, we have to recall some more facts about it. The logical skeleton of Calderón-Zygmund theory is the following.[7]

Assume we have a kernel $k(x)$ in $\mathbb{R}^n$ satisfying the "standard estimates" (3.1), (3.4) and the vanishing property (3.3). Let us consider the "truncated" operator

$$T_\varepsilon f(x) = \int_{\mathbb{R}^n} k_\varepsilon(x-y) f(y)\, dy$$

where the kernel k_ε is truncated near the origin, to cut out the nonintegrable singularity of k. If the truncation is done smoothly, the kernels k_ε will still satisfy assumptions (3.1), (3.4), (3.3), with the constants in (3.1), (3.4) independent of ε. So $T_\varepsilon f$ is well defined for, say, any $f \in C_0^\infty(\mathbb{R}^n)$. The first step of the theory consists in proving that the Fourier transform $\widehat{k_\varepsilon}$ is bounded, uniformly in ε; this proof exploits the three assumptions on k_ε, in particular the *vanishing property* is crucial. Plancherel theorem then implies that T_ε maps $L^2(\mathbb{R}^n)$ into itself continuously. Namely:

$$\|T_\varepsilon f\|_2 = \|k_\varepsilon * f\|_2 = \left\|\widehat{k_\varepsilon * f}\right\|_2 = \left\|\widehat{k_\varepsilon}\widehat{f}\right\|_2 \le \left\|\widehat{k_\varepsilon}\right\|_\infty \|\widehat{f}\|_2 \le c\, \|f\|_2 \,.$$

The second step consists in proving that T_ε satisfies a weak $(1, 1)$ estimate, which means that:

$$\left|\{x \in \mathbb{R}^n : |T_\varepsilon f(x)| > \lambda\}\right| \le c\,\frac{\|f\|_{L^1(\mathbb{R}^n)}}{\lambda} \quad \text{for any } \lambda > 0,\ f \in L^1(\mathbb{R}^n).$$

This proof only exploits the property (3.4) of the kernel, plus the fact, already proved, that T_ε is continuous on L^2. The proof of the second step is the more delicate part of the theory, and is based on very original ideas from real analysis. Once one knows that T_ε is strongly L^2 continuous and weakly $(1, 1)$ continuous, a classical interpolation result by Marzinkievic implies the continuity of T_ε on L^p for $1 < p < 2$. A duality argument, and the symmetry property of the kernel, imply L^p continuity also in the range $2 < p < \infty$. All these continuity estimates on T_ε hold with constants independent of ε. Finally, one has to prove the actual convergence of T_ε to some T, which as a consequence will satisfy the same L^p bound. This can be a delicate point in the abstract setting, but in applications to PDEs usually is not a problem, since the existence of the limit as $\varepsilon \to 0$ is a starting point (representation formulas).

---

[7] All the details can be found in the books [21, Chap. 2] or [20, Chap. 1].

In 1970 Coifman-De Guzmán [5] found that what we have called "the second step of the theory", namely the proof that an L^2 continuous singular integral satisfies a weak $(1, 1)$-estimate, can be broadly generalized. Namely, all the real analysis machinery which is involved in the proof can be rephrased in any *space of homogeneous type*, which means a set X endowed with a *quasidistance* d, that is a function satisfying the axioms of distance, with the triangle inequality replaced by the weaker

$$d(x, y) \leq c\,[d(x, z) + d(z, y)]\,\forall x, y, z \in X$$

for some $c \geq 1$, and a positive Borel measure μ satisfying the *doubling condition*:

$$\mu(B(x, 2r)) \leq c\mu(B(x, r))\,\forall x \in X, r > 0,$$

where the balls $B(x, r)$ are those defined by the quasidistance, and the topology is the one induced by these balls.[8] The theory of singular integrals in spaces of homogeneous type was firstly systematized in the book by Coifman-Weiss [6], 1971, which is usually considered the date of birth of the theory. Now, let us point out the following fact: the hardest part of the classical theory of singular integrals is (quite easily and naturally) generalized to spaces of homogeneous type, while it is the "simple" proof of L^2 continuity that, being based in the classical case on the Fourier transform, does not admit any easy extension. So, from the very beginning, the crux of all generalizations of singular integral theories has been the proof of L^2 estimates. The first general idea to prove L^2 estimates in some non-Euclidean context (hence without using the Fourier transform) was Coltlar's *almost orthogonality principle*, an explanation of which can be found for instance in [20, Chap. 7, Sect. 2]. Without going into details, let us say that this technique allows to prove L^2 continuity of a singular integral assuming, as a first step, some quite strong vanishing property of the kernel, analogous to (3.3); the second step consists in showing how to reduce to this particular case a more general one, when the kernel is only supposed to have, say, a *uniformly bounded* (but not necessary vanishing) integral over spherical shells. However, this program is very hard to be carried out in a general space of homogeneous type. To realize the kind of difficulty involved, note that in a general measure metric space one does not have a translation or a convolution; singular integral operators are defined as

$$P.V. \int_X k(x, y) f(y)\, d\mu(y)$$

and the kernel is seen as a function of *two* variables. Hence, the most obvious way to generalize the vanishing property amounts to asking

---

[8] Rigorously speaking, there is an annoying further assumption which must be done: one has to assume that the balls are open with respect to the topology they induce, a property which is not automatic if d is not a distance. However, in any reasonable example this assumption is fulfilled.

$$\int_{R_1<d(x,y)<R_2} k\,(x,y)\,d\mu\,(y) = 0 \;\forall R_2 > R_1 > 0, \forall x \in X, and$$

$$\int_{R_1<d(x,y)<R_2} k\,(y,x)\,d\mu\,(y) = 0 \;\forall R_2 > R_1 > 0, \forall x \in X.$$

Now, it is very hard to reduce the general case of a kernel satisfying some reasonable cancellation property to the particular case when *both* these conditions are simultaneously satisfied. Actually, a really general, flexible theory capable of getting L^2 continuity of singular integrals in spaces of homogeneous type has been built only in 1985, as a by-product of a deep advance in the study of the theory of singular integrals in Euclidean spaces: the so-called $T\,(b)$-theorem by David-Journé-Semmes [7]. The important L^2 result which was proved, instead, already in 1971, by Knapp-Stein [17], is settled in a less general framework, namely that of *homogeneous groups*, which is a natural generalization of the abstract structure we have in the Heisenberg group.

Definition 33 *(See [20, Chap. 13, Sect. 5]). Assume $\mathbb{R}^N$ is endowed with a Lie group structure, that is a group operation $\circ$ ("translation") such that $(x, y) \longmapsto x \circ y$ and $x \mapsto x^{-1}$ are C^∞ functions. It is not restrictive to assume that 0 is the identity. Also, assume we are given a family $\{D\,(\lambda)\}_{\lambda>0}$ of group automorphisms ("dilations") acting as*

$$D\,(\lambda)\,(x) = \left(\lambda^{\alpha_1}x_1, \lambda^{\alpha_2}x_1, ..., \lambda^{\alpha_N}x_N\right)$$

where $\alpha_1, \alpha_2, ..., \alpha_N$ are positive numbers. This structure is called homogeneous group. *The homogeneous dimension is by definition the number $Q = \alpha_1 + \alpha_2 + ... + \alpha_N$.*

In any homogeneous group one can define a homogeneous norm, that is a non-negative function $x \mapsto \|x\|$ satisfying properties (3.5), (3.6), (3.7).

In this context one can study singular integrals of convolution type, with homogeneous kernel. Under natural assumptions which are the exact translation of those which hold for classical Calderón-Zygmund kernels, namely (3.10), (3.11), (3.12), the L^2 continuity result can be actually established, applying the almost-orthogonality principle. A detailed proof of this fact, firstly contained in [17], can be found in [20, Chap. 13, Sect. 5.3].

Let us summarize. Around 1970 the theory of Calderón-Zygmund singular integrals was extended at two different levels of generality. The proof of L^2 continuity was achieved in *homogeneous groups* (translations + dilations), assuming a quite rich underlying structure; the proof of L^p continuity, $1 < p < \infty$ (for an operator already known to be L^2 continuous) was established in a much greater generality, namely those of *spaces of homogeneous type* (quasidistance + doubling measure). Although not completely satisfactory for this logical asymmetry, these theories were sufficient to achieve deep results in terms of L^p estimates for subelliptic equations, as we are going to explain.

3.2.6 L^p Estimates for the Sublaplacian and the Kohn-Laplacian on the Heisenberg Group

By now we have picked together all the ingredients to explain how Folland-Stein [12] prove their L^p estimates. Summarizing, the situation is the following. We consider the operator

$$\mathcal{L}_\alpha \equiv \sum_{j=1}^{n} \left(X_j^2 + Y_j^2 \right) + i\alpha T \text{ in } \mathbb{R}^{2n+1}$$

where $\alpha \in \mathbb{C}$ is any admissible value, that is $\pm\alpha \neq n, n+2, n+4, \ldots$ Then, by Theorem 30, there exists a homogeneous, translation invariant fundamental solution γ_α such that the representation formula (3.9) can be established. The singular kernel $X_k X_j \gamma_\alpha$ on the homogeneous group $\mathbb{H}^n$ satisfies the assumptions (3.10), (3.11), (3.12) which allow to apply the results by Knapp-Stein [17] and Coifman-Weiss [6] and conclude that the operator

$$g(\xi) \mapsto P.V. \int_{\mathbb{R}^{2n+1}} \left(X_k X_j \gamma_\alpha \right) \left(\eta^{-1} \circ \xi \right) g(\eta) \, d\eta$$

is L^2 continuous, and therefore L^p continuous for any $p \in (1, \infty)$. It follows

$$\left\| X_k X_j u \right\|_{L^p(\mathbb{R}^{2n+1})} \leq c_{p,\alpha} \left\| \mathcal{L}_\alpha u \right\|_{L^p(\mathbb{R}^{2n+1})} \text{ for any } u \in C_0^\infty \left(\mathbb{R}^{2n+1} \right).$$

An analogous estimate clearly holds for $Y_k Y_j u$, $Y_k X_j u$ and Tu. In particular, for $\alpha = 0$, we have the first result of L^p estimates for second order derivatives (with respect to the vector fields) for an operator "sum of squares of Hörmander's vector fields". Introducing the Sobolev spaces $S^{p,k}$ of functions having L^p derivatives of order k with respect to the vector fields X_k, Y_k, Folland-Stein [12] extend the previous result to the following:

$$\|u\|_{S^{k+2,p}(\mathbb{R}^{2n+1})} \leq c_{p,\alpha} \left\{ \|\mathcal{L}_\alpha u\|_{S^{k,p}(\mathbb{R}^{2n+1})} + \|u\|_{S^{k,p}(\mathbb{R}^{2n+1})} \right\}$$

for any nonnegative integer k. This extension is by no means straightforward; anyhow, here we skip any detail for the sake of brevity. We will say something about this problem in a more general context of homogeneous groups (see Sect. 3.3.5)

At last, one can come back to the Kohn-Laplacian. Recall that, by Theorem 24 we have, for any $q = 0, 1, 2, \ldots, n$,

$$\Box_b \left(\sum_{|J|=q} \phi_J d\bar{z}^J \right) = \sum_{|J|=q} (\mathcal{L}_\alpha \phi_J) \, d\bar{z}^J \text{ with } \alpha = n - 2q.$$

By the diagonal form of the operator $\Box_b$, the previous regularity estimates have an immediate counterpart in terms of $\Box_b$. Actually, Folland-Stein in [12] are interested in studying the Kohn-Laplacian on a more general setting than the Heisenberg group: they consider $\Box_b$ on "nondegenerate CR manifolds" (see [12, Part III]) and by a suitable approximation technique, they reduce the study of the operators $\mathcal{L}_\alpha$ on these manifolds to that of $\mathcal{L}_\alpha$ on $\mathbb{H}^n$. We will not go into details here, also because a similar idea, but in a more general context, will be explained in Sect. 3.4.

3.3 Hörmander's Operators on Homogeneous Groups

Here we want to discuss some ideas of the second fundamental paper which this chapter is about, namely the one by Folland, 1975, Arkiv für Mat., [9].

3.3.1 Homogeneous Groups

We have already defined what is a homogeneous group (see Sect. 3.2.5), but let us recall the definition here, to fix terminology in view of further definitions. For this material and for the proofs of some facts we refer to the paper [9] and to the book [20, Chap. 13, Sect. 5]; a much wider presentation of this theory can be found in the book [2]; this last book is particularly suggested to find a good wealth of explicit examples of homogeneous groups and corresponding sublaplacians, with computations carried out in detail.

We say that a *homogeneous group* $\mathbb{G}$ is $\mathbb{R}^N$ endowed with a Lie group structure such that the group operation is written $u \circ v$ and called *translation*; the inverse is denoted by u^{-1}, and the identity is the origin, 0, and a one parameter family of automorphisms, called *dilations* and denoted by $D(\lambda)$, which act as follows:

$$D(\lambda) : (u_1, \ldots, u_N) \mapsto \left(\lambda^{\alpha_1} u_1, \ldots, \lambda^{\alpha_N} u_N\right) \ \forall \lambda > 0$$

for suitable fixed exponents[9] $0 < \alpha_1 \leq \alpha_2 \leq \ldots \leq \alpha_N$. The number

$$Q = \sum_{i=1}^{N} \alpha_i$$

---

[9] If $D(\lambda)$ is a family of dilations defined by an N-tuple of positive real numbers $(\alpha_1, \alpha_2, ..., \alpha_N)$, then for any $k > 0$ also the exponents $(k\alpha_1, k\alpha_2, ..., k\alpha_N)$ will define a family of dilations. Hence we can always normalize the exponents so that $\alpha_1 = 1$. We will see later that, under the further assumptions that we will make, if α_1 is an integer then all the other exponents are integers.

is called the *homogeneous dimension*[10] of $\mathbb{G}$. If $\varphi : \left(\mathbb{R}^N, \circ \right) \to \left(\mathbb{R}^N, * \right)$ is any group isomorphism, we can also say that

$$v = \varphi(u)$$

is another choice of a system of coordinates in $\mathbb{G}$.

Example 34 *(1) Heisenberg group* $\mathbb{H}^1$ *in* $\mathbb{R}^3$:

$$(x_1, y_1, t_1) \circ (x_2, y_2, t_2) = (x_1 + x_2, y_1 + y_2, t_1 + t_2 + 2(x_2 y_1 - x_1 y_2))$$
$$D(\lambda)(x, y, t) = \left(\lambda x, \lambda y, \lambda^2 t \right)$$

(2) Kolmogorov-type group in $\mathbb{R}^3$:

$$(x_1, y_1, t_1) \circ (x_2, y_2, t_2) = (x_1 + x_2, y_1 + y_2 - x_1 t_2, t_1 + t_2)$$
$$D(\lambda)(x, y, t) = \left(\lambda x, \lambda^3 y, \lambda^2 t \right).$$

More examples will be given in Sect. 3.3.6. While one can define a great variety of smooth group laws in $\mathbb{R}^N$, the requirement of the existence of "dilations" is a strict limitation with important consequences. The following can be proved:

Proposition 35 *(See [20, pp. 618–9]) If $\mathbb{G}$ is a homogeneous group then the group law $P(x, y) = x \circ y$ is polynomial. More precisely,*

$$x \circ y = x + y + Q(x, y) \tag{3.13}$$

with

$$Q(0, 0) = Q(x, 0) = Q(0, y) = 0.$$

Actually, Q does not contain pure monomials in x or y; all its terms are mixed and their homogeneous degree (in the usual Euclidean sense) is at least 2. Writing $Q = (Q_1, Q_2, ..., Q_N)$, each Q_k is α_k-homogeneous, that is

$$Q_k(D(\lambda) x, D(\lambda) y) = \lambda^{\alpha_k} Q_k(x, y) \,\forall \lambda > 0.$$

Also, the Jacobian matrix of $x \longmapsto P(x, y)$ is triangular, with ones along the diagonal and the same is true for $y \longmapsto P(x, y)$. Therefore the Lebesgue measure is both left and right invariant with respect to the group translations. We say that dx is the Haar measure of $\mathbb{G}$.

Finally, it is possible to choose a system of coordinates in $\mathbb{G}$ such that $u^{-1} = -u$.

---

[10] For the reason explained in the previous footnote, in actual applications of the theory this number will be an integer.

We can summarize some of the previous properties in terms of useful *rules for changing variables inside an integral*:

$$
\begin{aligned}
x \circ y = x' &\implies dx = dx' \\
y \circ x = x' &\implies dx = dx' \\
x^{-1} = x' &\implies dx = dx' \\
x = D(\lambda) x' &\implies dx = \lambda^{Q} dx'.
\end{aligned}
$$

In any homogeneous group $\mathbb{G}$ we can define a *homogeneous norm*, that is a function $\|\cdot\| : \mathbb{G} \to \mathbb{R}$, smooth outside the origin and such that: $\|u\| = 0 \iff u = 0$; $\|D(\lambda)u\| = \lambda \|u\|$ for every $u \in \mathbb{G}$, $\lambda > 0$;

there exists $c(\mathbb{G}) \geq 1$ such that for every $u, v \in \mathbb{G}$

$$
\|u \circ v\| \leq c \left(\|u\| + \|v\| \right).
$$

There are several concrete ways to define a homogeneous norm (all of them being equivalent), among the most used we quote:

$$
(1) \quad \|u\| = \left(\sum_{k=1}^{N} |u_k|^{Q/\alpha_k} \right)^{1/Q}
$$

$$
(2) \quad \|u\| = \max_{k=1,2,\ldots,N} |u_k|^{1/\alpha_k}
$$

$$
(3) \quad \|u\| = \rho \quad \Leftrightarrow \quad \left| D\left(\frac{1}{\rho} \right) u \right| = 1
$$

for any $u \in \mathbb{G}$, $u \neq 0$, where $|\cdot|$ denotes the Euclidean norm and $\|0\| = 0$.

The homogeneous norm (3) has the special property that the set $\{u \in \mathbb{G} : \|u\| = 1\}$ coincides with the Euclidean unit sphere Σ_N, which is sometimes useful.

Recall that we are assuming $u^{-1} = -u$ (see the above proposition); this implies $\|u^{-1}\| = \|u\|$ for any of the three homogeneous norms above (while in general one has $\|u^{-1}\| \leq c \|u\|$). We will always assume in the following $\|u^{-1}\| = \|u\|$.

Defining

$$
d(u, v) = \left\| v^{-1} \circ u \right\|
$$

we get that d is a *quasidistance*:

$$
d(u, v) \geq 0 \text{ and } d(u, v) = 0 \text{ if and only if } u = v;
$$

$$
d(u, v) = d(v, u)
$$

$$
d(u, v) \leq c \{ d(u, z) + d(z, v) \}
$$

for every $u, v, z \in \mathbb{R}^N$ and some positive constant $c(\mathbb{G}) \geq 1$.

If we denote by $B(u, r) \equiv B_r(u) \equiv \{v \in \mathbb{R}^N : d(u, v) < r\}$ the metric balls, then we see that $B(0, r) = D(r)B(0, 1)$. Moreover, since the Lebesgue measure in $\mathbb{R}^N$ is the Haar measure of $\mathbb{G}$,

$$|B(u, r)| = |B(0, 1)| \, r^Q,$$

for every $u \in \mathbb{G}$ and $r > 0$, where Q is the homogeneous dimension of $\mathbb{G}$. Hence, we see that a homogeneous group is a particular *space of homogeneous type* in the sense of Coifman-Weiss (see §[6]).

The *convolution* of two functions in $\mathbb{G}$ is defined as

$$(f * g)(x) = \int_{\mathbb{R}^N} f(x \circ y^{-1}) \, g(y) \, dy = \int_{\mathbb{R}^N} g(y^{-1} \circ x) \, f(y) \, dy, \qquad (3.14)$$

for every couple of functions for which the above integrals make sense. Note that this convolution *is not commutative*. If P is any *left invariant differential operator*,

$$P(f * g) = f * Pg \qquad (3.15)$$

(provided the integrals converge). Note that, differently from the Euclidean case, we cannot interchangeably take the differential operator P on f or g. Observe that, like P is left invariant with respect to translation, (3.15) says that it is left invariant with respect to convolution. (Actually, to remember the positions of the variables in definition (3.14), one can think that it is given this way just in order for (3.15) to be true).

We say that *a differential operator P on $\mathbb{G}$ is homogeneous of degree $\delta > 0$* if

$$P\left(f\left(D(\lambda)x\right)\right) = \lambda^\delta \, (Pf)(D(\lambda)x)$$

for every test function f, $\lambda > 0$, $x \in \mathbb{R}^N$. Also, we say that *a function f is homogeneous of degree $\delta \in \mathbb{R}$* if

$$f\left(D(\lambda)x\right) = \lambda^\delta \, f(x) \text{for every } \lambda > 0, \, x \in \mathbb{R}^N.$$

Clearly, if P is a differential operator homogeneous of degree δ_1 and f is a homogeneous function of degree δ_2, then Pf is homogeneous of degree $\delta_2 - \delta_1$. For example, $x_i \frac{\partial}{\partial x_j}$ is homogeneous of degree $\alpha_j - \alpha_i$.

3.3.2 Homogeneous Lie Algebras

Let us consider now the *Lie algebra ℓ* associated to the group $\mathbb{G}$, that is, the Lie algebra of left-invariant vector fields, endowed with the Lie bracket given by the commutator $[X, Y]$ of vector fields. We can fix a basis $X_1, \ldots, X_N$ in ℓ choosing X_i

as the left invariant vector field which agrees with $\frac{\partial}{\partial x_i}$ at the origin. Explicitly, we can compute:

$$X_j f(x) = \partial_{y_j} [f(x \circ y)]_{/y=0}.$$

This identity holds because: at $x = 0$ it is true, since X_j coincides with ∂_j; and the vector field $f \mapsto \frac{\partial}{\partial y_j} [f(x \circ y)]_{/y=0}$ is readily seen to be left invariant; hence it must coincide with X_j, which is the only left invariant vector field which agrees with ∂_j at the origin.

Example 36 *On the Heisenberg group* $\mathbb{H}^1$*, we have*

$$(x, y, t) \circ (x', y', t') = (x + x', y + y', t + t' + 2(x'y - xy')).$$

Hence the vector field which coincides with ∂_x at the origin can be computed as follows:

$$\partial_{x'} \left[f(x + x', y + y', t + t' + 2(x'y - xy')) \right]_{/(x',y',t')=(0,0,0)}$$
$$= \partial_x f(x, y, t) + 2y \partial_t f(x, y, t)$$

and

$$X = \partial_x + 2y \partial_t$$

(as we already know).

By (3.13) we then have

$$X_j f(x) = \partial_{y_j} [f(x + y + Q(x, y))]_{/y=0} = \partial_{x_j} f + \sum_{j<k} q_j^k(x) \partial_{x_k} f \qquad (3.16)$$

where $q_j^k(x)$ is homogeneous of degree $\alpha_k - \alpha_j$. In particular, X_j is α_j-homogeneous.
Therefore, we can also extend the dilations $D(\lambda)$ to ℓ setting

$$D(\lambda) X_i = \lambda^{\alpha_i} X_i.$$

Then $D(\lambda)$ turns out to be a Lie algebra automorphism, i.e.,

$$D(\lambda) [X, Y] = [D(\lambda)X, D(\lambda)Y].$$

In this sense, ℓ is said to be a *homogeneous Lie algebra*; as a consequence, ℓ is *nilpotent* (see [20, pp. 621–622]): there exists an integer k such that any commutator of length greater than k vanishes.

It is worthwhile to note that, by the triangular form (3.16) of the vectors fields, on any homogeneous group the transposed of X_j is simply $-X_j$ (while for a vector field we usually have $X_j^* = -X_j + f_j$).

3.3.3 Hörmander's Operators on Homogeneous Groups

So far, we have introduced notions and discussed properties which hold in *any* homogeneous group.

We are now going to concentrate on two special classes of homogeneous groups, and corresponding Lie algebras, for which a left invariant, 2-homogeneous Hörmander's operator exists. Although from an abstract point of view these are just particular cases, the following development of the theory has shown that these classes of left invariant homogeneous operators are extremely useful to attack the local study of *any* Hörmander's operator.

In the following we keep the previous notation and assumptions: we have a homogeneous group $\mathbb{G}$ in $\mathbb{R}^N$ and the left invariant vector fields $X_1, X_2, ..., X_N$ which coincide at the origin with $\partial_{x_1}, \partial_{x_2}, ...\partial_{x_N}$. We are going to make further assumptions on these X_i's.

Assumption A. Among the vector fields $\{X_i\}_{i=1}^N$ there exist $X_1, X_2, ..., X_q$ $(q < N)$ which are 1-homogeneous and satisfy Hörmander's condition.

Note that the full system $\{X_i\}_{i=1}^N$ trivially satisfies Hörmander's condition (without necessity of taking commutators!), since they are independent at the origin, *and therefore at any point*, by left invariance. So assumption A consists in asking the existence of a "small" number of generators, *all being 1-homogeneous*.

Under assumption A the sum of squares operator

$$L = \sum_{i=1}^q X_i^2$$

is left invariant, 2-homogeneous and hypoelliptic, by Hörmander's theorem. Also, note that by the remark at the end of Sect. 3.3.2, the transposed of L is L itself. The simplest example of this kind of operator is the sublaplacian on the Heisenberg group. Another consequence of Assumption A is that the Lie algebra ℓ has a particular simple structure. If we define:

$$V_1 = \text{span}\left(X_1, X_2, ..., X_q\right);$$
$$V_2 = [V_1, V_1];$$
$$...$$
$$V_{k+1} = [V_k, V_1];$$
$$...$$

then, by the homogeneity of the Lie algebra, we see that each vector field in V_j is j-homogeneous[11]; this implies that $V_j \cap V_k = \{0\}$ for $j \neq k$; moreover, since the

---

[11] In particular this means that in this case the homogeneity exponents α_j, and therefore the homogeneous dimension Q, are integers.

$\{X_i\}_{i=1}^q$ satisfy Hörmander's condition, there exists an integer r such that

$$\ell = V_1 \oplus V_2 \oplus ... \oplus V_r \tag{3.17}$$

(as a vector space decomposition), and $V_k = \{0\}$ for $k > r$. We say that *the Lie algebra ℓ is stratified*, V_j is its j-th *layer*, and the *generators of the first layer*, $X_1, X_2, ..., X_q$, satisfy Hörmander's condition at step r. Explicitly:

Definition 37 *A Lie algebra is stratified if (3.17) holds for suitable subspaces V_k satisfying*

$$[V_1, V_k] = V_{k+1} \, for \, k = 1, 2, ..., r - 1$$
$$[V_1, V_r] = \{0\}.$$

Definition 38 *A homogeneous group $\mathbb{G}$ in $\mathbb{R}^N$ such that the vector fields X_i's defined as above satisfy "Assumption A" is called* stratified group *or* (homogeneous) Carnot group. *If $X_1, X_2, ..., X_q$ are generators of the first layer, the (hypoelliptic, left invariant, 2-homogeneous) operator*

$$L = \sum_{i=1}^q X_i^2$$

is called sublaplacian *on the group $\mathbb{G}$.*

The analysis of sublaplacians on stratified groups has become a very broad field of research. We point out the (already quoted) monograph by Bonfiglioli, Lanconelli, Uguzzoni [2], 2007, which is entirely devoted to this.

Alternatively to Assumption A, we will consider the situation described by the following:

Assumption B. Among the vector fields $\{X_i\}_{i=1}^N$ there exist $X_0, X_1, X_2, ..., X_q$ $(q + 1 < N)$ satisfying Hörmander's condition[12] such that $X_1, X_2, ..., X_q$ are 1-homogeneous while X_0 is 2-homogeneous.

Under assumption B the operator

$$L = \sum_{i=1}^q X_i^2 + X_0$$

is left invariant, 2-homogeneous and hypoelliptic, by Hörmander's theorem. Moreover,

---

[12] This X_0 is actually one of the initial vector fields $\{X_i\}_{i=1}^N$, which just for convenience we relabel X_0.

$$L^* = \sum_{i=1}^{q} X_i^2 - X_0$$

is still hypoelliptic.

Example 39 *The simplest example of this kind of operator is Kolmogorov operator:* $\mathbb{G} = \mathbb{R}^3$ *with the following translations and dilations:*

$$(x_1, y_1, t_1) \circ (x_2, y_2, t_2) = (x_1 + x_2, y_1 + y_2 - x_1 t_2, t_1 + t_2)$$
$$D(\lambda)(x, y, t) = \left(\lambda x, \lambda^3 y, \lambda^2 t\right).$$

As to the vector fields,

$$X_1 = \partial_x;$$
$$X_0 = x\partial_y + \partial_t$$

and the operator is

$$L = X_1^2 + X_0 = \partial_x^2 + x\partial_y + \partial_t.$$

Under Assumption B, the Lie algebra ℓ is not necessarily stratified (its structure is sometimes called "stratified of type II"), while it is always *graded*, a notion which however is not necessary to explain here (see [9] for details).

3.3.4 Homogeneous Fundamental Solutions and L^p Estimates

At this point we recall the fundamental result by Folland [9]:

Theorem 40 *(Existence of a homogeneous fundamental solution) Let $\mathcal{L}$ be a left invariant differential operator homogeneous of degree two on a homogeneous group $\mathbb{G}$, such that $\mathcal{L}$ and $\mathcal{L}^*$ are both hypoelliptic. Moreover, assume $Q \geq 3$ (where Q is the homogeneous dimension of $\mathbb{G}$). Then there is a unique translation invariant fundamental solution Γ (that is, $\Gamma(x, y) = \Gamma(y^{-1} \circ x)$) such that:*

(a) $\Gamma \in C^\infty\left(\mathbb{R}^N \setminus \{0\}\right)$;
(b) Γ is homogeneous of degree $(2 - Q)$;
(c) for every distribution τ,

$$\mathcal{L}(\tau * \Gamma) = (\mathcal{L}\tau) * \Gamma = \tau. \tag{3.18}$$

The result proved by Folland is actually more general: he assumes $\mathcal{L}$ homogeneous of degree $\alpha > 0$ and $Q > \alpha$. We will not be interested in the case $\alpha \neq 2$. Note that the restriction $Q > 2$ is irrelevant, since the only case it excludes is that of elliptic equations in two variables with constant coefficients.

This theorem in particular applies to the left invariant 2-homogeneous Hörmander's operators on $\mathbb{G}$ that we have considered under Assumption A or B, namely

$$\mathcal{L} = \sum_{i=1}^{q} X_i^2 \text{ in case A}$$

$$\mathcal{L} = \sum_{i=1}^{q} X_i^2 + X_0 \text{ in case B,}$$

and from now on we will always consider an operator $\mathcal{L}$ of one of these types. However, note that in the statement of the theorem nothing is required about the *explicit form* of the operator $\mathcal{L}$.

Let us explicitly note the relevance of the information contained in the *two* identities (3.18). Namely, saying that Γ is a fundamental solution for $\mathcal{L}$ means that

$$\mathcal{L}\Gamma = \delta, \text{ which implies}$$
$$\mathcal{L}(\tau * \Gamma) = \tau * \mathcal{L}\Gamma = \tau * \delta = \tau.$$

However, due to the noncommutativity of the convolution, we cannot write, in general $\mathcal{L}(f * g) = \mathcal{L}f * g$, hence the identity $\mathcal{L}(\tau * \Gamma) = (\mathcal{L}\tau) * \Gamma$ is not trivial, but is actually a further information which is supplied by the theorem, which implies that we can represent any distribution τ as $\tau = (\mathcal{L}\tau) * \Gamma$.

The proof of the above theorem is nonconstructive: first, applying some deep abstract results from distribution theory, Folland shows the existence of a "local fundamental solution" smooth outside the pole; then, starting from this kernel and exploiting the dilations on $\mathbb{G}$ and the homogeneity of $\mathcal{L}$, he builds a new kernel which also possesses the desired homogeneity. In the end, we do not have any idea of the explicit form of this Γ. However, its abstract properties are really enough to make the theory work. To begin with, one can prove, as in the case dealt by Folland-Stein [12] for the Heisenberg group, all the properties which are collected in the following theorem:

Theorem 41 *The following representation formula holds, for any test function u:*

$$X_i X_j u = PV\left((\mathcal{L}u) * X_i X_j \Gamma\right) + c_{ij} \mathcal{L}u \text{ for } i, j = 1, 2, ..., q$$

where c_{ij} are constants. Explicitly:

$$X_i X_j u(x) = \lim_{\varepsilon \to 0} \int_{\|y^{-1} \circ x\| > \varepsilon} X_i X_j \Gamma(y^{-1} \circ x)(\mathcal{L}u)(y)\, dy + c_{ij} \mathcal{L}u(x).$$

Moreover, the kernel $K(x) = X_i X_j \Gamma(x)$ satisfies the following properties:

$$|K(x)| \le \frac{c}{\|x\|^Q};$$

$$. \ |K(x \circ y) - K(x)| + |K(y \circ x) - K(x)| \le c\frac{\|y\|}{\|x\|^{Q+1}}$$

whenever $\|x\| \ge 2\|y\|$;

$$\int_{R_1 < \|x\| < R_2} K(x)\,dx = \int_{R_1 < \|x\| < R_2} K\left(x^{-1}\right)\,dx = 0 \text{ for any } 0 < R_1 < R_2 < \infty.$$

In case B, one also has, for any test function u:

$$X_0 u = PV((\mathcal{L}u) * X_0 \Gamma) + c_0 \mathcal{L}u$$

where c_0 is a constants; the kernel $K = X_0\Gamma$ satisfies the same assumptions of the kernel K above.

In view of the previous theorem, the same singular integral machinery used by Folland-Stein in [12] applies to the present situation (recall that Knapp-Stein L^2 result holds in any homogeneous group and Coifman-Weiss weak $(1, 1)$ result holds in any space of homogeneous type) and allows to say that:

$$\left\|X_i X_j u\right\|_{L^p(\mathbb{R}^N)} \le c\,\|\mathcal{L}u\|_{L^p(\mathbb{R}^N)} \quad \text{for } i, j = 1, 2, ..., q,$$

any test function u, $1 < p < \infty$, and an analogous estimate on $\|X_0 u\|_{L^p(\mathbb{R}^N)}$ holds in case B.

3.3.5 Higher Order Estimates

The previous discussion could suggest the wrong idea that, after all, the desired L^p estimates follow by general abstract results, without making real computations with the vector fields. In order to correct this unfair feeling we are now going to discuss, although rather briefly, the way how higher order estimates are proved. Namely, we would like to know that whenever $\mathcal{L}u$ possesses X_j-derivatives of order k, then u possesses X_j-derivatives of order $k + 2$. This is actually the case, at least under assumption A, but is far from being obvious.

The idea is that, starting from the representation formula

$$X_i X_j u = PV\left((\mathcal{L}u) * X_i X_j \Gamma\right) + c_{ij}\mathcal{L}u$$

we would like to take, in the convolution, one X_i-derivative from $X_i X_j \Gamma$ to $\mathcal{L}u$, getting something like

$$X_i X_j u = (X_i \mathcal{L}u) * X_j \Gamma + c_{ij}\mathcal{L}u. \tag{3.19}$$

If this were true, we could then perform another X_k-derivative, getting

$$X_k X_i X_j u = PV\left((X_i \mathcal{L}u) * X_k X_j \Gamma\right) + c_{kj} X_i \mathcal{L}u + c_{ij} X_k \mathcal{L}u. \tag{3.20}$$

and then

$$\left\| X_k X_i X_j u \right\|_p \le c \left\| X \mathcal{L}u \right\|, \tag{3.21}$$

which is the first iterative step to get the desired result. However, the noncommutativity of convolution on $\mathbb{G}$ prevents us from writing

$$(\mathcal{L}u) * X_i X_j \Gamma = X_i (\mathcal{L}u) * X_j \Gamma$$

because on a homogeneous group the identity

$$(Xf) * g = f * Xg \text{ is false.}$$

What can be proved, instead, is the following identity

$$(X_i f) * g = f * X_i^R g \tag{3.22}$$

where X_i^R is the right-invariant vector field with coincides at the origin with ∂_{x_i}, and hence with X_i.

Since both the systems $\{X_i\}$ and $\{X_i^R\}$ have the following "triangular form" with respect to Cartesian derivatives:

$$X_i = \partial_{x_i} + \sum_{k=i+1}^{N} q_i^k(x)\, \partial_{x_k}$$

$$X_i^R = \partial_{x_i} + \sum_{k=i+1}^{N} \overline{q}_i^k(x)\, \partial_{x_k}$$

where $q_i^k, \overline{q}_i^k$ are polynomials, homogeneous of degree $\alpha_k - \alpha_i$, we can see that every X_i can be rewritten in terms of the X_k^R, namely

$$X_i u = \sum_{k=1}^{N} X_k^R \left(\beta_{ik}(x)\, u\right)$$

with β_{ik} are homogeneous of degree $\alpha_i - \alpha_k$. Then, we rewrite

$$X_i X_j u = PV\left((\mathcal{L}u) * X_i X_j \Gamma\right) + c_{ij}\mathcal{L}u$$

$$= PV\left((\mathcal{L}u) * \sum_{k=1}^{N} X_k^R \left(\beta_{ik} X_j \Gamma\right)\right) + c_{ij}\mathcal{L}u.$$

Next,

$$PV\left((\mathcal{L}u) * \sum_{k=1}^{N} X_k^R \left(\beta_{ik} X_j \Gamma\right)\right)$$

$$= \sum_{k=1}^{q} PV\left((\mathcal{L}u) * X_k^R \left(\beta_{ik} X_j \Gamma\right)\right) + \sum_{k=q+1}^{N} PV\left((\mathcal{L}u) * X_k^R \left(\beta_{ik} X_j \Gamma\right)\right).$$

In the first sum we can apply (3.22) getting

$$\sum_{k=1}^{q} PV\left((\mathcal{L}u) * X_k^R \left(\beta_{ik} X_j \Gamma\right)\right) = \sum_{k=1}^{q} X_k (\mathcal{L}u) * \left(\beta_{ik} X_j \Gamma\right)$$

with β_{ik} constants, which is similar to the form (3.19) that we hoped to get.

In the second sum we cannot proceed the same way because the X_k derivative for $k > q$ has the "weight" of a higher order derivative. However, any such X_k^R can be expressed as a linear combination of commutators of the $X_1^R, X_2^R, ...X_q^R$ of the first layer, so that we can always reduce to the previous situation. Note that, doing so, we arrive to a representation formula of the kind

$$X_i X_j u = \sum_{h} X_h \mathcal{L}u * K_h + c_{ij}\mathcal{L}u$$

where K_h is a kernel $(1 - Q)$-homogeneous, which however is not simply one of the functions $X_k \Gamma$ (for $k = 1, 2, ..., q$). Taking one more X_k derivative of both sides we get something which is similar to (3.20), and therefore produces an L^p estimate of type (3.21).

3.3.6 Some Classes of Examples of Homogeneous Groups and Corresponding Hörmander's Operators

In this paragraph we collect some explicit examples of homogeneous groups which satisfy Folland's assumptions. All of them are extracted from the book [2, Chaps. 3, 4], where further details can be found.

Homogeneous Groups of Heisenberg Type (or H-Groups). These groups are a generalization of the Heisenberg groups $\mathbb{H}^n$ that we have already described. Consider the Lie group in $\mathbb{R}^{m+n} \ni (x, t)$:

$$(x, t) \circ (\xi, \tau) = \left(x + \xi, t_1 + \tau_1 + \frac{1}{2} \left\langle B^{(1)} x, \xi \right\rangle, ..., t_n + \tau_n + \frac{1}{2} \left\langle B^{(n)} x, \xi \right\rangle \right)$$

where $B^{(1)}, ..., B^{(n)}$ are constant $m \times m$ skew symmetric orthogonal matrices, also satisfying $B^{(i)} B^{(j)} = -B^{(j)} B^{(i)}$ for every $i \neq j$. With this translation $\circ$ and the dilations

$$D (\lambda) (x, t) = \left(\lambda x, \lambda^2 t \right),$$

$\mathbb{R}^{m+n}$ becomes a homogeneous group of step two, called (prototype) group of Heisenberg type, or (prototype) H-group. The sublaplacian on this group is

$$L = \sum_{i=1}^m X_i^2 \text{ with}$$

$$X_i = \partial_{x_i} + \frac{1}{2} \sum_{k=1}^n \left(\sum_{l=1}^m b_{i,l}^{(k)} x_l \right) \partial_{t_k}; \ i = 1, 2, ..., m$$

where $B^{(k)} = \left(b_{i,l}^{(k)} \right)_{i,l=1}^m$.

For these sublaplacian an explicit fundamental solution is known:

$$\Gamma (x, t) = c \left(|x|^4 + 16 |t|^2 \right)^{\frac{2-Q}{4}} \text{ with } Q = m + 2n.$$

These groups and operators have been introduced by Kaplan [16] in 1980. See [2, Chap. 3] for more information about them.

Kolmogorov Type Groups. In $\mathbb{R}^{1+N} \ni (t, x)$ let us define the composition law:

$$(t, x) \circ (t', x') = (t + t', x + E (t') x)$$
$$(t, x)^{-1} = (-t, -E (-t) x)$$

with

$$E (t') = \exp (t' B),$$

and B is the $N \times N$ triangular matrix with the following structure:

$$B = \begin{bmatrix} 0 & 0 & \dots & 0 & 0 \\ B_1 & 0 & \dots & 0 & 0 \\ 0 & B_2 & \dots & \dots & \dots \\ \dots & \dots & \dots & 0 & 0 \\ 0 & 0 & \dots & B_r & 0 \end{bmatrix}$$

where B_j is a $p_j \times p_{j-1}$ block with rank equal to p_j, for every $j = 1, 2, \dots, r$. Moreover $p_0 \geq p_1 \geq \dots \geq p_r$ and $p_0 + p_1 + \dots + p_r = N$. Splitting

$$\mathbb{R}^N = \mathbb{R}^{p_0} \times \mathbb{R}^{p_1} \times \dots \times \mathbb{R}^{p_r} \ni \left(x^{(0)}, x^{(1)}, \dots, x^{(r)} \right),$$

one sees that this group is homogeneous with the dilations:

$$D(\lambda)(t, x) = \left(\lambda t, \lambda x^{(0)}, \lambda^2 x^{(1)}, \dots, \lambda^{r+1} x^{(r)} \right).$$

The sublaplacian in this case is

$$L = \sum_{j=1}^{p_0} \partial_{x_j}^2 + Y^2 \text{ with}$$

$$Y = \partial_t + \langle Bx, \nabla_x \rangle.$$

These operators have been introduced by Lanconelli-Polidoro [18] in 1994, and generalize the operator studied by Kolmogorov in 1934. See [2, Chap. 4]. We will say more about these operators in Chap. 5.

An explicit example in the above class is:

$$B = \begin{bmatrix} 0 & 0 & 0 \\ 1 & 0 & 0 \\ 0 & 1 & 0 \end{bmatrix}; N = p_0 + p_1 + p_2 = 3; \exp(sB) = \begin{bmatrix} 1 & 0 & 0 \\ s & 1 & 0 \\ \frac{s^2}{2} & s & 1 \end{bmatrix}$$

$$(t, x_1, x_2, x_3) \circ (s, y_1, y_2, y_3) = \left(t + s, x_1 + y_1, x_2 + y_2 + sx_1, x_3 + y_3 + sx_2 + \frac{s^2}{2} x_1 \right)$$

$$D(\lambda)(t, x_1, x_2, x_3) = \left(\lambda t, \lambda x_1, \lambda^2 x_2, \lambda^3 x_3 \right)$$

$$L = \partial_{x_1}^2 + \left(\partial_t + x_1 \partial_{x_2} + x_2 \partial_{x_3} \right)^2$$

This class of examples can be naturally modified to give examples of operators satisfying "Assumption B". With the same translation, the operator

$$L = \sum_{j=1}^{p_0} \partial_{x_j}^2 + Y \text{ with}$$

$$Y = \partial_t + \langle Bx, \nabla_x \rangle$$

is left invariant and 2-homogeneous with respect to the dilations:

$$D(\lambda)(t, x) = \left(\lambda^2 t, \lambda x^{(0)}, \lambda^3 x^{(1)}, \lambda^5 x^{(2)}, ..., \lambda^{2r+1} x^{(r)}\right).$$

We will come back to these operators in Chap. 5. Actually, this is the original form in which these operators have been studied in [18], as Kolmogorov-Fokker-Planck operators with linear drift.

Heat Type Operators. Let $\mathbb{G} = \left(\mathbb{R}^N, \circ, D(\lambda)\right)$ be a homogeneous group of homogeneous dimension Q and let

$$L = \sum_{i=1}^{q} X_i^2$$

be a corresponding sublaplacian. Let us define the homogeneous group

$$\mathbb{H} = \left(\mathbb{R}^{N+1}, *, D'(\lambda)\right)$$

such that:

$$(x, t) * (y, s) = (x \circ y, t + s)$$

$$D'(\lambda)(x, t) = \left(D(\lambda) x, \lambda^2 t\right).$$

The above structure actually defines a homogeneous group, of homogeneous dimension $Q' = Q + 2$; the operators:

$$H_{\mp} = L \pm \partial_t$$

are 2-homogeneous, left invariant (in $\mathbb{H}$) Hörmander's operators ("backward" and "forward" heat operators), possessing, by Folland's theorem, a $2 - Q'$ (that is, $-Q$) homogeneous fundamental solution ("heat kernel"). The fundamental solution $h(t, x)$ of $L - \partial_t$ vanishes for $t < 0$. Since, by the hypoellipticity of $L - \partial_t$, $h(t, x)$ must be C^∞ outside the pole, it has to vanish of infinite order as $t \to 0^+$ with $x \neq 0$. This argument however does not imply the validity of specific Gaussian estimates, which one can expect from such heat kernels. The theme of Gaussian estimates for heat-type Hörmander's operator will be addressed in the middle 1980's.

Bony-Type Sublaplacians. Let us consider the following Hörmander's operator in $\mathbb{R}^{2+N} \ni (t, s, x)$:

$$L = T^2 + S^2 \text{ with}$$

$$T = \partial_t; \, S = \partial_s + t\partial_{x_1} + \frac{t^2}{2}\partial_{x_2} + \dots + \frac{t^N}{N!}\partial_{x_N}$$

which is 2-homogeneous with respect to the dilations

$$D(\lambda)(t, s, x) = \left(\lambda t, \lambda s, \lambda^2 x_1, \lambda^3 x_2, \dots, \lambda^{N+1} x_N\right)$$

and left invariant with respect to the translations:

$$(t, s, x) \circ (\tau, \sigma, y) = \left(t + \tau, \sigma + s, x_1 + y_1 + \sigma t, \dots, x_N + y_N + \sum_{k=1}^{N} \frac{y_k t^{N-k}}{(N-k)!} + \frac{\sigma t^N}{N!}\right).$$

For more informations, see [2, Chap. 4].

3.4 General Hörmander's Operators

We finally come to the last and more general paper of the three we are discussing, namely the one by Rothschild-Stein, 1976, Acta Math. [19].

3.4.1 The Problem, and How to Approach It

Here the issue is to prove L^p estimates on $X_i X_j u$ of the same kind already proved on homogeneous groups, in the general situation where such underlying group structure is lacking. Let us recall two examples of Hörmander's operators which do not admit an underlying homogeneous group:

(i) The Grushin operator:

$$L = \partial_{xx}^2 + x\partial_y. \tag{3.23}$$

Here the absence of an underlying group of translation is made evident by the fact that at points (x_0, y_0) with $x_0 \neq 0$ the vector fields $X_1 = \partial_x$, $X_0 = x\partial_y$ are independent, while if $x_0 = 0$ they are not. In contrast with this, if two left invariant vector fields are independent at one point, they must be independent everywhere. The reason is that if $c_1 X_1 + c_0 X_0 = 0$ at some point, with c_1, c_2 constants, then the vector field $Y = c_1 X_1 + c_0 X_0$ is also left invariant and vanishes at some point; this implies that it vanishes everywhere. (A left invariant vector field is *uniquely determined* by its value at one point).

(ii) The Mumford operator:

$$L = \partial_t + \cos\theta\,\partial_x + \sin\theta\,\partial_y + \frac{\sigma^2}{2}\partial_{\theta\theta}^2.$$

Here the absence of a family of dilations is made evident by the fact that the coefficients $\cos\theta$, $\sin\theta$ are not polynomials.

The idea of attacking the study of general Hörmander's vector fields making use of the theory already developed in the case of homogeneous groups had been explicitly declared in the paper [9] by Folland, as a motivation for the development of that theory, and its roots date back to the paper [12] by Folland-Stein, where this idea had been already developed in a particular case:

> In the theory of *elliptic* operators the constant-coefficients operators serve as a useful class of models for the general situation: constant-coefficients operators are amenable to treatment by the techniques of Euclidean harmonic analysis (Fourier transforms, convolution operators, etc.), and the results obtained thereby can usually be extended to the variable-coefficient case by perturbation arguments. Now, a constant-coefficient operator is nothing more than a translation-invariant operator on the Abelian Lie group $\mathbb{R}^N$. From this point of view, it is natural to attempt to construct a class of models for non-elliptic operators on certain non-Abelian Lie groups. The Lie algebras of the groups involved should have a structure which reflects the behavior of the commutators in the original problem and the groups themselves should admit a "harmonic analysis" which will produce results similar to those of the Euclidean case. A particular case of this program has been carried out in considerable detail in Folland-Stein [12, 13], in which sharp L^p and Lipschitz (or Hölder) estimates for the $\overline{\partial}_b$ complex on the boundary of a complex domain with nondegenerate Levi form are obtained by using certain left-invariant operators in the Heisenberg group as models. (see [9, p. 162]).

Now, what does it mean to handle a general Hörmander's operator by a 2-homogeneous left invariant one "by a perturbation argument"? Let us give a simple example:

Example 42 *The vector fields*

$$X = \partial_x + 2y\partial_t;\ Y = \partial_y - 2\left(e^x - 1\right)\partial_t$$

satisfy Hörmander's condition at step 2; the coefficients of Y are not polynomials (hence cannot be left invariant with respect to any structure of homogeneous group), but if we take the first order expansion near $x = 0$, $e^x - 1 \sim x$, we get the vector fields

$$X = \partial_x + 2y\partial_t;\ Y' = \partial_y - 2x\partial_t$$

which are 1-homogeneous and left invariant on the Heisenberg group $\mathbb{H}^1$. We note that:

(1) $X, Y', \left[X, Y'\right]$ are three independent vectors at any point, exactly like X, Y, $[X, Y]$. In other words, Y and Y' are indistinguishable from the point of view of

the Lie algebra structure, up to the step which is required to check Hörmander's condition.

(2) Moreover, near $x = 0$ we have

$$Y - Y' = -2\left(e^x - 1 - x\right)\partial_t.$$

The interesting property of this vector field is that, with respect to the homogeneities of the Heisenberg group (x has weight 1, t has weight 2), $Y - Y'$ is "approximately homogeneous of order 0 near $x = 0$", since $-2\left(e^x - 1 - x\right) \sim -x^2$ and $-x^2\partial_t$ is homogeneous of degree 0, while Y' is homogeneous of degree 1. This implies that if Γ is, say, a $(2 - Q)$-homogeneous function, $Y'\Gamma$ will be $(1 - Q)$-homogeneous, that is more singular than Γ, while $\left(Y - Y'\right)\Gamma$ will be approximately $(2 - Q)$-homogeneous, that is as singular as Γ. As we will see later in more detail, this is the key point which makes this approximation useful.

Generalizing the above example, the idea is to take the Taylor polynomials of some fixed order, in a neighborhood of some point x_0, of the coefficients of the vector fields X_i and to approximate each X_i with the corresponding Y_i having polynomial coefficients. More precisely, if the original vector fields satisfy Hörmander's condition at step r, this means that not more than $r - 1$ derivatives of the coefficients need to be computed when checking this condition, hence the vector fields obtained replacing each coefficient with its Taylor polynomial of degree $r - 1$ will satisfy the same relevant commutator relations. Actually, the Lie algebras generated by the two sets of vector fields will be *indistinguishable up to step r*.

Now, we would like that the polynomial vector fields Y_i were 2-homogeneous and left invariant with respect to some structure of homogeneous group. However we know that, on the one hand, the Lie algebras of the X_i's and the Y_i's have the same structure up to step r, while, on the other hand, the Lie algebra generated by a system of homogeneous left invariant vector field always possesses some special property: for instance, its structure must be the same at any point. But this means that we cannot hope to approximate our original system of vector fields with a "good" one unless our original system *already satisfies some additional algebraic condition* which, in particular, makes its Lie algebra "of constant structure". We could express more precisely this condition saying that for any choice of N vector fields, among $X_1, X_2, ..., X_q$ and their commutators up to step r, if these vectors are independent at *some* point then they must be independent at *any* point.

Now, Rothschild-Stein's idea is *to add extra variables to our original vector fields*, in order to fulfil this condition.

Example 43 *Let us consider Grushin's vector fields*

$$X_1 = \partial_x, X_0 = x\partial_y \text{ which "live" in } \mathbb{R}^2$$

and generate a Lie algebra which has not the same structure at any point, because X_1, X_0 *are independent if and only if* $x \neq 0$. *Starting with these vector fields, we can build the new ones*

$$\widetilde{X}_1 = \partial_x, \widetilde{X}_0 = X_0 + \partial_t = x\partial_y + \partial_t \text{ which "live" in } \mathbb{R}^3.$$

Note that $\widetilde{X}_1, \widetilde{X}_0, [\widetilde{X}_1, \widetilde{X}_0]$ *are independent at any point of* $\mathbb{R}^3$. *Their Lie algebra is the same as that of the Heisenberg group* $\mathbb{H}^1$, *and actually a smooth change of variables in* $\mathbb{R}^3$ *can turn these vector fields into the "canonical form"* $X' = \partial_{x'} + 2y'\partial_{t'}, Y' = \partial_{y'} - 2x'\partial_{t'}$ *of* $\mathbb{H}^1$. *The vector fields* $\widetilde{X}_1, \widetilde{X}_0$ *satisfy the desired condition, moreover they project onto the original* X_1, X_2, *in the sense that for any function* $f(x, y)$ *which does not depend on the added* t *variable, we have*

$$X_1 f = \widetilde{X}_1 f; X_0 f = \widetilde{X}_0 f.$$

This property should make easy to get the desired a priori estimates for $X_i X_j u$ *once we have proved analogous estimates for* $\widetilde{X}_i \widetilde{X}_j u$ *in a higher dimensional space.*

We say that the vector fields X_1, X_2 have been *lifted* to $\widetilde{X}_1, \widetilde{X}_2$. In the above simple example, the lifted vector fields are already left invariant and homogeneous. More generally one expects to build up a two-step process:

Example 44 *Let us consider the operator*

$$L = X_1^2 + X_2^2 \text{ with}$$
$$X_1 = \partial_x, X_2 = (e^x - 1)\partial_y \text{ in } \mathbb{R}^2.$$

(The vector fields have the same structure than Grushin's vector fields, but nonpolynomial coefficients). Then:
  First step: we lift X_1, X_2 *to*

$$\widetilde{X}_1 = \partial_x, \widetilde{X}_2 = (e^x - 1)\partial_y + \partial_t \text{ in } \mathbb{R}^3.$$

Note that $\widetilde{X}_1, \widetilde{X}_2, [\widetilde{X}_1, \widetilde{X}_2]$ *are independent at any point of* $\mathbb{R}^3$.
  Second step: we approximate $\widetilde{X}_1, \widetilde{X}_2$ *with*

$$Y_1 = \partial_x, Y_2 = x\partial_y + \partial_t$$

which are left invariant and 1-homogeneous in $\mathbb{H}^1$ *(up to a smooth change of coordinates).*

3.4.2 Lifting

It is time to reformulate the previous discussion giving a precise definition and stating a theorem.

Definition 45 *We say that a system of smooth vector fields* $Z_1, Z_2, ..., Z_q$ *is free up to step* r *in a domain* Ω *of* $\mathbb{R}^N$ *if the* Z_i's *and their commutators up to step* r *do not satisfy any linear relation other than those which hold automatically as a consequence of antisymmetry of the Lie bracket and Jacobi identity.*

To put it into another way, $Z_1, Z_2, ..., Z_q$ are free up to step r if and only if the only relations of linear dependence which we can write among them and their commutators up to step r (at any point of Ω), are those which can be established *without knowing the coefficients* of the Z_j's.

If the vector fields satisfy Hörmander's condition at step r and are free up to step r, then in particular their Lie algebra has "constant structure". As we will see with the examples, however, the converse may not be true.

Example 46 *1. The "usual" vector fields* X, Y *on* $\mathbb{H}^1$ *are free up to step 2:*

$$X, Y, [X, Y]$$

are linearly independent, and there are not other commutators to be considered, up to step 2.
2. The Grushin vector fields

$$X = \partial_x, Y = x\partial_y$$

are not free at step 2 because

$$Y = x\,[X, Y]$$

which is a nontrivial relation between a generator and a commutator of step 2. In this case, as already noted, the choice of a natural basis of $\mathbb{R}^2$ *is different from point to point.*
3. The "usual" vector fields X_1, X_2, Y_1, Y_2 *on* $\mathbb{H}^2$ *are not free up to step 2, because*

$$[X_1, Y_1] = [X_2, Y_2]$$

and this is a nontrivial relation between two commutators of step 2. Note that in this case the Lie algebra actually has the same structure at any point.

We can now state the lifting theorem, the first famous result proved by Rothschild-Stein in [19]. To simplify the language, we state it only for operators "sum of squares", that is we take $X_0 = 0$.

Theorem 47 *(Lifting, see [19, Thm. 4]) Let $X_1, ..., X_q$ be vector fields satisfying Hörmander's condition of step r at x_0. (This clearly implies that such property holds in a suitable neighborhood of x_0). Then there exist an integer m and vector fields $\widetilde{X}_k$ defined in a neighborhood of $(x_0, 0) \in \mathbb{R}^{n+m}$, of the form*

$$\widetilde{X}_k = X_k + \sum_{j=1}^{m} u_{kj} \left(x, t_1, t_2, ..., t_{j-1} \right) \partial_{t_j}$$

($k = 1, ..., q$), where the u_{kj}'s are polynomials, such that the $\widetilde{X}_k$'s are free up to step r and satisfy Hörmander's condition at step r in this neighborhood. (Meaning that the $\widetilde{X}_k$ and their commutators up to step r span $\mathbb{R}^{n+m}$).

The proof of this theorem is long and very technical, and we will not say anything more about it.

3.4.3 Approximation with Left Invariant Vector Fields

Starting from a system of vector fields $X_1, X_2, ..., X_q$ which satisfy Hörmander's condition at step r in a neighborhood of $x_0 \in \mathbb{R}^n$, we have therefore built a new family of "lifted" vector fields $\widetilde{X}_1, \widetilde{X}_2, ..., \widetilde{X}_q$, which are *free up to step r* and satisfy Hörmander's condition at step r in a neighborhood of $(x_0, 0) \subset \mathbb{R}^{n+m}$. The second step of the theory consists in approximating these free vector fields with homogeneous left invariant vector fields on a suitable homogeneous group.

To approach this problem, we start with some algebraic remarks. Disregarding the explicit form of the vector fields $\widetilde{X}_k$ in Cartesian coordinates, the structure of their Lie algebra up to step r (that is, the number and type of independent objects among the $\widetilde{X}_k$'s and their commutators up to step r) is completely determined by the requirement of being *free up to step r*. This also means that the dimension $N = n+m$ of the lifted space only depends on the numbers q and r.

Example 48 *If $q = 2$ and $r = 3$ we have to consider the following independent vector fields:*

$$\widetilde{X}_1, \widetilde{X}_2; \left[\widetilde{X}_1, \widetilde{X}_2 \right]; \left[\widetilde{X}_1, \left[\widetilde{X}_1, \widetilde{X}_2 \right] \right]; \left[\widetilde{X}_2, \left[\widetilde{X}_1, \widetilde{X}_2 \right] \right].$$

Since these vector fields are 5, this means that $N = 5$. We actually don't know wether the higher order commutators, like

$$\left[\widetilde{X}_1, \left[\widetilde{X}_1, \left[\widetilde{X}_1, \widetilde{X}_2 \right] \right] \right]$$

vanish, or satisfy other nontrivial relations (this depends on the actual explicit form of the vector fields in Cartesian coordinates). However, up to step 3, their algebra is determined by the numbers $q = 2$ and $r = 3$.

Now the idea is that the Lie algebra of homogeneous left invariant vector fields Y_k which approximate locally the $\tilde{X}_k$'s can be defined abstractly as the free Lie algebra of step r on q generators,[13] which turns out to be a homogeneous Lie algebra; this means that the vector fields Y_k and their commutators up to step r satisfy exactly the same relations than the $\tilde{X}_k$'s, but moreover all their commutators of step $> r$ vanish. The structure of homogeneous group $\mathbb{G}$ in $\mathbb{R}^N$ is, in turn, determined by that of the corresponding Lie algebra. The vector fields $\tilde{X}_k$ are defined in a neighborhood of $\xi_0 = (x_0, 0) \in \mathbb{R}^N$; the vector fields Y_k are defined in the whole $\mathbb{R}^N$, and their behavior near the origin will approximate the behavior of the $\tilde{X}_k$ near ξ_0. This means that the approximation between $\tilde{X}_k$ and Y_k is realized *in a suitable system of coordinates*. The precise approximation result proved by Rothschild-Stein is the following:

Theorem 49 *(Approximation, see [19, Thm. 5]) Assume $\tilde{X}_1, \tilde{X}_2, ..., \tilde{X}_q$, are free up to step r and satisfy Hörmander's condition at step r in a neighborhood of $(x_0, 0) \in \mathbb{R}^{n+m}$. There exist a structure of homogeneous group $\mathbb{G}$ on $\mathbb{R}^N$, $N = n+m$, a family of homogeneous left invariant Hörmander's vector fields $Y_1, Y_2, ..., Y_q$ on $\mathbb{G}$ and, for any η in a neighborhood of $(x_0, 0)$, a smooth diffeomorphism*

$$\xi \mapsto \Theta_\eta(\xi)$$

from a neighborhood of η onto a neighborhood of the origin in $\mathbb{G}$, smoothly depending on η, such that for any smooth function $f : \mathbb{G} \to \mathbb{R}$,

$$\tilde{X}_i \left(f \left(\Theta_\eta \left(\cdot \right) \right) \right) (\xi) = \left(Y_i f + R_i^\eta f \right) \left(\Theta_\eta (\xi) \right) \tag{3.24}$$

($i = 1, ..., q$) where the "remainder" R_i^η is a smooth vector fields of local degree ≤ 0 and smoothly depends on the parameter η.

The previous statement needs to be completed by the following

Definition 50 *A differential operator P on $\mathbb{G}$ is said to have* local degree less than or equal to k *if, after taking the Taylor expansion at 0 of its coefficients, each term obtained is homogeneous of degree $\leq k$.*

Also, the role of $\Theta_\eta(\xi)$ is better explained by the following

Theorem 51 *The function*

$$\rho(\xi, \eta) = \left\| \Theta_\eta(\xi) \right\|$$

is a quasidistance (in the neighborhood where it is defined). Its relation with the Euclidean one is expressed by

$$c_1 |\xi - \eta| \leq \rho(\xi, \eta) \leq c_2 |\xi - \eta|^{1/r}$$

---

[13] The free Lie algebra of step r on q generators is defined as the quotient of the free Lie algebra on q generators with the ideal spanned by the commutators of length at least $r + 1$.

where r is the step at which Hörmander's condition holds. Denoting the ρ-balls with the symbol $B(\xi, R)$ we have

$$c_1 R^Q \leq |B(\xi, R)| \leq c_2 R^Q$$

for R small enough. In particular, the Lebesgue measure is (locally) doubling.

Remark 52 *We encounter here for the first time a notion of (quasi) distance which is related to a system of Hörmander's vector fields which are not translation invariant. The simple behavior of the volume of metric balls in this case heavily depends on the fact that the vector fields $\widetilde{X}_i$ are free. As we will see in Sect. 4.2, the geometry related to a general system of Hörmander's vector fields can be much more involved.*

In order to better understand in which sense the vector field R_i^η in (3.24) can be seen as a "small remainder", let us consider the action of $\widetilde{X}_i$ on a function $f(\Theta_\eta(\xi))$ when f is homogeneous of some negative degree $-\alpha$ on $\mathbb{G}$ and smooth outside the origin. In this case we have

$$Y_i f \quad \text{is} \quad (-\alpha - 1)\text{-homogeneous, hence}$$

$$\left| Y_i f(\Theta_\eta(\xi)) \right| \leq \frac{c}{\left\| \Theta_\eta(\xi) \right\|^{\alpha+1}} = \frac{c}{\rho(\xi, \eta)^{\alpha+1}},$$

while $R_i^\eta f$ is not a homogeneous function, but nevertheless, since R_i^η has local degree ≤ 0,

$$\left| R_i^\eta f(\Theta_\eta(\xi)) \right| \leq \frac{c}{\left\| \Theta_\eta(\xi) \right\|^{\alpha}} = \frac{c}{\rho(\xi, \eta)^{\alpha}}$$

that is: *this term is less singular than $Y_i f$.*

The three theorems we have just stated (lifting; approximation; properties of the map Θ), together with some more properties of the map $\Theta_\eta(\xi)$ proved in [19] and which we will recall when appropriate, constitute a powerful tool, known as "Rothschild-Stein's lifting and approximation technique", which has proved to have further applications than the one for which it was originally designed, and has actually been used by several authors.[14] The general idea is that, thanks to this technique, the study of local properties of a general Hörmander's operator can sometimes be reduced to the study of a homogeneous left invariant Hörmander's operator, of the kind studied by Folland in [9]. Also, sometimes it is useful the first result alone (lifting theorem) to reduce the study of a general system of Hörmander's vector fields to a system of *free* (although not homogeneous) Hörmander's vector fields, which has some advantages.

Since the original proof of the lifting and approximation theorem given in [19] is long and difficult, several authors have given alternative proofs: Hörmander-Melin

---

[14] We will illustrate some examples of this fact in Chap. 4.

[15], Folland [10] and Goodman [14] prove the lifting theorem and a pointwise version of the approximation theorem, without dealing with the map $\Theta_\eta(\cdot)$ and its properties; Folland restricts to the particular case when the starting vector fields are already left invariant and homogeneous with respect to a group structure (but are not free); Bonfiglioli-Uguzzoni [1] have proved that under Folland's assumptions, the original vector fields can be lifted directly to free left invariant homogeneous vector fields (in other words, in this case the "remainders" R_i^η can be taken equal to zero). In the case of general Hörmander's vector fields, Christ-Nagel-Stein-Wainger [4] prove a more general version of the lifting theorem, because they consider "weighted" vector fields.

We now come on illustrating the way this machinery is used in proving a priori L^p estimates for general Hörmander's operators.

3.4.4 Parametrix and L^p Estimates

We will concentrate on the case of an operator "sum of squares of Hörmander's vector fields" (satisfying Hörmander's condition at step r),

$$L = \sum_{i=1}^q X_i^2$$

defined and satisfying these assumptions in a neighborhood of some $x_0 \in \mathbb{R}^n$.

We consider the corresponding lifted operator

$$\widetilde{L} = \sum_{i=1}^q \widetilde{X}_i^2$$

where the $\left\{\widetilde{X}_i\right\}_{i=1}^q$ are free up to step r and satisfy Hörmander's condition at step r in some neighborhood $\widetilde{\Omega} = \Omega \times I$ of $(x_0, 0) \in \mathbb{R}^{n+m} = \mathbb{R}^N$ (Thm. 47). Let's note that if we are able to prove L^p estimates for $\widetilde{L}$, that is

$$\|u\|_{S^{2,p}(\widetilde{\Omega}')} \le c \left\{ \|\widetilde{L}u\|_{L^p(\widetilde{\Omega})} + \|u\|_{L^p(\widetilde{\Omega})} \right\} \tag{3.25}$$

for $1 < p < \infty$, $\widetilde{\Omega}' \Subset \widetilde{\Omega}$, $u \in C^\infty(\widetilde{\Omega})$, where

$$\|u\|_{S^{2,p}} = \sum_{i,j=1}^q \left\|\widetilde{X}_i\widetilde{X}_j u\right\|_{L^p} + \sum_{k=1}^q \left\|\widetilde{X}_k u\right\|_{L^p} + \|u\|_{L^p}$$

then we immediately get similar estimates for L, that is the original "unlifted operators". Namely, applying the previous estimate to

$$u(x, t) = f(x)$$

with $f \in C^\infty(\Omega)$ and recalling that $X_i f = \widetilde{X}_i f$, we get

$$\|f\|_{S^{2,p}(\Omega')} \le c \left\{ \|Lf\|_{L^p(\Omega)} + \|f\|_{L^p(\Omega)} \right\}.$$

Hence, it is enough to prove (3.25) in the lifted space. Let us consider the structure of homogeneous group $\mathbb{G}$ in $\mathbb{R}^N$, the corresponding 1-homogeneous left invariant vector fields Y_i in $\mathbb{G}$ and the map $\Theta_\eta(\xi)$, defined for ξ, η in a neighborhood of $\xi_0 = (x_0, 0)$, as appear in Thm. 49. The operator

$$\mathcal{L} = \sum_{i=1}^{q} Y_i^2$$

is a 2-homogeneous, left invariant, Hörmander's operator on $\mathbb{G}$. By Folland's theory [9], it admits a $(2 - Q)$-homogeneous, left invariant, fundamental solution Γ smooth outside the pole, such that

$$\mathcal{L} \int_{\mathbb{R}^N} \Gamma\left(v^{-1} \circ u\right) f(v)\, dv = f(u)$$

for every $f \in C_0^\infty(\mathbb{R}^N)$. Starting with Γ, Rothschild-Stein build a *parametrix* for $\widetilde{L}$. Let us consider the kernel

$$\Gamma\left(\Theta_\eta(\xi)\right)$$

and let us compute

$$\widetilde{L}\left[\Gamma\left(\Theta_\eta(\cdot)\right)\right](\xi).$$

Recall that, by Thm. 49,

$$\widetilde{X}_i\left[\Gamma\left(\Theta_\eta(\cdot)\right)\right](\xi) = \left(Y_i\Gamma + R_i^\eta\Gamma\right)\left(\Theta_\eta(\xi)\right),$$

hence

$$\widetilde{L}\left[\Gamma\left(\Theta_\eta(\cdot)\right)\right](\xi) = (\mathcal{L}\Gamma)\left(\Theta_\eta(\xi)\right) + \sum_{i=1}^{q}\left(Y_i R_i^\eta\Gamma + Y_i R_i^\eta\Gamma + R_i^\eta R_i^\eta\Gamma\right)\left(\Theta_\eta(\xi)\right)$$

$$= \delta_0\left(\left(\Theta_\eta(\xi)\right)\right) + \left(R^\eta\Gamma\right)\left(\Theta_\eta(\xi)\right)$$

where δ_0 is the Dirac mass and R^η has local degree ≤ 1, therefore

$$\left|\left(R^\eta\Gamma\right)\left(\Theta_\eta(\xi)\right)\right| \le \frac{c}{\left\|\Theta_\eta(\xi)\right\|^{Q-1}} = \frac{c}{\rho(\xi, \eta)^{Q-1}}$$

that is, $(R^\eta \Gamma)(\Theta_\eta(\xi))$ behaves like a *fractional* (nonsingular) integral kernel. The above computation contains the basic idea. More precisely, we need to define the parametrix as

$$K(\xi, \eta) = a(\xi) \Gamma(\Theta_\eta(\xi)) b(\eta)$$

with a, b cutoff functions, with support small enough so that $K(\xi, \eta)$ is defined on the whole space (recall that $\Theta_\eta(\xi)$ is only locally defined). Then we define the integral operator

$$Pf(\xi) = \int K(\xi, \eta) f(\eta) \, d\eta$$

and compute, for any test function f, $\widetilde{L}(Pf)$, finding (by the above computation)

$$\widetilde{L}(Pf)(\xi) = a(\xi) f(\xi) + \int K_1(\xi, \eta) f(\eta) \, d\eta \qquad (3.26)$$

where K_1 is a suitable nonsingular (locally integrable) kernel, satisfying a growth estimate like

$$|K_1(\xi, \eta)| \le \frac{c}{\rho(\xi, \eta)^{Q-1}}.$$

Since we would like to have a representation formula of f in terms of $\widetilde{L} f$, modulo a nonsingular integral operator, we now *transpose* the identity (3.26), finding

$$P^*\left(\widetilde{L}^* f\right)(\xi) = a(\xi) f(\xi) + \int K_1(\eta, \xi) f(\eta) \, d\eta.$$

Recall that $\widetilde{L}^* = \widetilde{L} + $ (lower order terms). In our sketch of Rothschild-Stein's argument we decide to ignore this point and simply write $\widetilde{L}^* = \widetilde{L}$, hence

$$a(\xi) f(\xi) = P^*\left(\widetilde{L} f\right)(\xi) - \int K_1(\eta, \xi) f(\eta) \, d\eta \qquad (3.27)$$

where (exchanging the variables ξ, η in the kernel of P),

$$P^* f(\xi) = \int a(\eta) \Gamma(\Theta_\xi(\eta)) b(\xi) f(\eta) \, d\eta.$$

We are ready to take two derivatives $\widetilde{X}_i \widetilde{X}_j$ of both sides of (3.27). For any ξ in a small ball where $a \equiv 1$, we have

$$\widetilde{X}_i \widetilde{X}_j f(\xi) = \widetilde{X}_i \widetilde{X}_j P^*\left(\widetilde{L} f\right)(\xi) - \widetilde{X}_i \widetilde{X}_j \int K_1(\eta, \xi) f(\eta) \, d\eta \qquad (3.28)$$

$$\equiv I + II.$$

Let us reflect on the two terms at the right-hand side. The term I is the second derivative of an integral operator whose kernel behaves like the fundamental solution Γ: morally speaking, we should get a singular integral; and this is actually the case: exploiting again the Approximation Theorem 49, one can prove

$$\widetilde{X}_i \widetilde{X}_j P^* \left(\widetilde{L} f \right) (\xi) = P.V. \int a(\eta) \left(Y_i Y_j \Gamma \right) \left(\Theta_\xi (\eta) \right) b(\xi) f(\eta) \, d\eta$$
$$+ c_{ij} (\xi) \left(\widetilde{L} f \right) (\xi) + (\text{remainders}) . \qquad (3.29)$$

The term II is the second derivative of a fractional integral operator whose kernel behaves like $\rho(\xi, \eta)^{1-Q}$; this means that *one* derivative of the integral operator should be a singular integral applied to f. To speed up the reasoning about these operators, some terminology must be introduced. Let us call:

operator of type 0 a singular integral operator;

operator of type 1 an integral operator whose kernel behaves like $\rho(\xi, \eta)^{1-Q}$ (hence, it can be differentiated once, before becoming a singular integral operator).

With this terminology, denoting operators of type k with a subscript k, we have

$$II = \widetilde{X}_i \widetilde{X}_j (S_1 f) = \widetilde{X}_i (S_0 f) .$$

The expression $\widetilde{X}_i (S_0 f)$ is an unpleasant one. However, Rothschild-Stein are able to prove that

$$\widetilde{X}_j (S_1 f) = \sum_{h,k} S_1^{(h)} \left(\widetilde{X}_k f \right) + S_1' f \qquad (3.30)$$

for suitable operators $S_1^{(h)}$, S_1' of type 1. Proving this relation of "commutation" between vector fields and integral operators is a heavy part of the theory, involving long and subtle reasonings.[15] Anyhow, once (3.30) is known, one readily has

$$II = \widetilde{X}_i \widetilde{X}_j (S_1 f) = \widetilde{X}_i \left[\sum_{h,k} S_1^{(h)} \left(\widetilde{X}_k f \right) + S_1' f \right]$$
$$= \sum_{h,k} S_0^{(h)} \left(\widetilde{X}_k f \right) + S_0' f. \qquad (3.31)$$

This, together with (3.28) and (3.29) shows that

$$\widetilde{X}_i \widetilde{X}_j f (\xi) = S_0 \left(\widetilde{L} f \right) (\xi) + c_{ij} (\xi) \left(\widetilde{L} f \right) (\xi) + \sum_{h,k} S_0^{(h)} \left(\widetilde{X}_k f \right) + S_0' f$$

Finally, taking L^p norms of both sides we get

---

[15] Just to give a glimpse of the kind of difficulty, we can say that the problem is analogous to the one described in Sect. 3.3.5 dealing with homogeneous groups, but in a more abstract context.

$$\left\| \tilde{X}_i \tilde{X}_j f \right\|_p \leq c \left\{ \left\| S_0 \left(\tilde{L} f \right) \right\|_p + \left\| \tilde{L} f \left(\xi \right) \right\|_p + \sum_{h,k} \left\| S_0^{(h)} \left(\tilde{X}_k f \right) \right\|_p + \left\| S_0' f \right\|_p \right\}.$$

Now, suppose we know that operators of type 0 map L^p into L^p continuously, then we would conclude

$$\left\| \tilde{X}_i \tilde{X}_j f \right\|_p \leq c \left\{ \left\| \tilde{L} f \right\|_p + \sum_k \left\| \tilde{X}_k f \right\|_p + \left\| f \right\|_p \right\}$$

and, introducing the Sobolev norms $S^{k,p}$ (with respect to vector fields),

$$\| f \|_{S^{2,p}(B_r)} \leq c \left\{ \| \tilde{L} f \|_{L^p(B_{2r})} + \| f \|_{S^{1,p}(B_{2r})} + \| f \|_{L^p(B_{2r})} \right\}$$

which is (almost) the desired estimate.

3.4.5 Singular Integral Estimates

So we are left to check that operators of type 0 are L^p continuous. Recall that these integral operators do not "live" on a homogeneous group but on a (local) space of homogeneous type, however a particularly simple one, whose structure can be seen as a small local perturbation of that of $\mathbb{G}$, via the diffeomorphism $u = \Theta_\xi (\eta)$. Namely, recall that:

$$\rho (\xi, \eta) = \left\| \Theta_\xi (\eta) \right\|$$

where $\|u\|$ is the homogeneous norm on $\mathbb{G}$;

$$|B (\xi, r)| \simeq r^Q \text{ for small } r;$$

moreover, the change of variables

$$u = \Theta_\xi (\eta)$$

within an integral, for η fixed, gives

$$d\xi = c (\eta) (1 + O (\|u\|)) \, du$$

with $c (\eta)$ smooth and bounded away from zero.

Just to give an idea of how these properties simplify the study of operators of type 0, let us check the cancellation property for the kernel

$$Y_i Y_j \Gamma \left(\Theta_\eta (\xi) \right).$$

Let us compute

$$\int_{R_1 < \rho(\xi, \eta) < R_2} Y_i Y_j \Gamma\left(\Theta_\eta\left(\xi\right)\right) d\xi$$

letting $u = \Theta_\xi\left(\eta\right)$

$$= c\left(\eta\right) \int_{R_1 < \|u\| < R_2} Y_i Y_j \Gamma\left(u\right)\left(1 + O\left(\|u\|\right)\right) du$$

$$= c\left(\eta\right) \int_{R_1 < \|u\| < R_2} Y_i Y_j \Gamma\left(u\right) du + c\left(\eta\right) \int_{R_1 < \|u\| < R_2} Y_i Y_j \Gamma\left(u\right) O\left(\|u\|\right) du$$

$$= A + B.$$

Now $A = 0$ by the vanishing property which holds in view of Folland's study of the homogeneous fundamental solution Γ on homogeneous groups, while

$$\left| Y_i Y_j \Gamma\left(u\right) O\left(\|u\|\right) \right| \le \frac{c}{\|u\|^{Q-1}},$$

hence a standard computation shows that

$$|B| \le c$$

uniformly in R_1, R_2. Hence we also have

$$\left| \int_{R_1 < \rho(\xi, \eta) < R_2} Y_i Y_j \Gamma\left(\Theta_\eta\left(\xi\right)\right) d\xi \right| \le c$$

for any $R_2 > R_1 > 0$.

The singular integral theory already used by Folland in [9], that is Coifman-Weiss [6] results coupled with those by Knapp-Stein [17], are still adaptable to this situation,[16] and L^p continuity of operators of type 0 can be actually proved. This concludes our account of Rothschild-Stein's paper.

Remark 53 *Let us point out some points in the paper by Rothschild-Stein which motivate further research in this field:*

1. *In contrast with the result proved in the case of homogeneous groups, here a priori estimates are only* local. *Proving L^p estimates for $X_i X_j u$ on the whole $\mathbb{R}^N$ for general Hörmander's operators is a difficult problem; we will say something more about this in the Chap. 5.*
2. *A priori estimates are stated, in the paper, with the language of regularity results: "if Lu belongs to this space, then u belongs to that space", and not actually*

---

[16] We have written that the theory is *adaptable*, not *directly applicable* to our situation. Rothschild-Stein [19] just sketch a proof of this adaptation. By now, however, this point is clearly established on the basis of more recent theories of singular integrals.

writing a priori estimates. If one writes down the results in terms of a priori estimates, one finds something like:

$$\|f\|_{S^{2,p}(B_r)} \le c\left\{\|Lf\|_{L^p(B_{2r})} + \|f\|_{S^{1,p}(B_{2r})} + \|f\|_{L^p(B_{2r})}\right\}.$$

Now, "taking to the left-hand side" the term $\|f\|_{S^{1,p}(B_{2r})}$ is not a trivial task with these Sobolev spaces. This issue has been addressed and answered many years later.

3. No explicit statement is done about the dependence of the constants. For instance, one could ask: if, for a system $(X_i)_{i=1}^q$ of Hörmander's vector fields, we consider an operator

$$L_A = \sum_{i,j=1}^{q} a_{ij} X_i X_j$$

where $A = (a_{ij})_{i,j=1}^q$ is a constant, symmetric, positive matrix, so that L_A can be actually rewritten as a Hörmander's operator, can we say that a-priori estimates hold with some constants depending on the matrix A only through the minimum and maximum eigenvalues? The answer cannot easily be read from the paper [19].

3.5 Some Final Comments on the Quest of a-Priori Estimates in Sobolev Spaces

3.5.1 Local Versus Global Estimates

The estimates proved both in homogeneous groups and in the general situation apply to test functions, hence are local in nature. If one applies them to a non-compactly supported function, what is got is the following:

$$\|f\|_{S^{2,p}(B_r)} \le c\left\{\|Lf\|_{L^p(B_{2r})} + \|f\|_{S^{1,p}(B_{2r})} + \|f\|_{L^p(B_{2r})}\right\}.$$

Apart from the problem of "taking to the left-hand side" the $S^{1,p}$ norm, these estimates are essentially local: starting with these, one can hope to get something like

$$\|f\|_{S^{2,p}(\Omega')} \le c\left\{\|Lf\|_{L^p(\Omega)} + \|f\|_{L^p(\Omega)}\right\} \text{ for } \Omega' \Subset \Omega,$$

but not for $\Omega' = \Omega$. This poses the question of proving estimates *near the boundary* in Sobolev norms: however, this is a completely open, difficult, problem.

3.5.2 Levels of Generality

The three papers we have discussed in this chapter, namely Folland-Stein [12], Folland [9], Rothschild-Stein [19], besides containing fundamental ideas, results and techniques which are currently used in the research on this field, are also representative of three different levels of generality in the study of Hörmander's operators, which still characterize the current research:

(1) sublaplacians on Heisenberg groups;
(2) Hörmander's operators on homogeneous groups;
(3) general Hörmander's operators.

Each of these environments has its open problems and challenges. Moreover, there are typical differences in the kind of results which are usually proved in different contexts. For instance, in homogeneous groups one hopes to prove global results, while for general systems of Hörmander's vector fields one usually confines to local results.

There is also a finer scale of generality which appears in some lines of research: from the less to the more general context one can study Hörmander's vector fields on:

- the Heisenberg group $\mathbb{H}^1$;
- Heisenberg groups $\mathbb{H}^n$;
- groups of Heisenberg type (also called H-groups);
- stratified groups of step 2;
- stratified (Carnot) groups;
- homogeneous (graded) Lie groups;
- Lie groups with polynomial growth;
- general Lie groups;
- general Hörmander's vector fields.

The possible requirement that the vector fields be free is a further assumption which can be made both in the context of groups and in the general situation.

Another important difference in generality, which is transversal with respect to the previous hierarchy, consists in studying sum of squares of Hörmander's vector fields *or* Hörmander's operators (containing a drift X_0). The possible presence of the drift X_0 can create deep additional problems. Nevertheless, we recall that Kolmogorov-Fokker-Planck operators, which constitute an important motivation for this theory (see Sect. 2.1), actually contain the drift.

So far, the literature devoted to general Hörmander's vector fields is considerably narrower than that dealing with Hörmander's vector fields on some kind of Lie group, as the literature devoted to Hörmander's operators (with drift) is considerably narrower than that dealing with sum of squares.

References

1. Bonfiglioli, A., Uguzzoni, F.: Families of diffeomorphic sub-Laplacians and free Carnot groups. Forum Math. **16**(3), 403–415 (2004)
2. Bonfiglioli, A., Lanconelli, E., Uguzzoni, F.: Stratified Lie Groups and Potential Theory for their Sub-Laplacians, Springer Monographs in Mathematics. Springer, Berlin (2007)
3. Calderón, A.P., Zygmund, A.: On the existence of certain singular integrals. Acta Math. **88**, 85–139 (1952)
4. Christ, M., Nagel, E., Stein, M., Wainger, S.: Singular and maximal Radon transforms: analysis and geometry. Ann. Math. **150**(2), 489–577 (1999)
5. Coifman, R.R., de Guzmán, M.: Singular integrals and multipliers on homogeneous spaces. Collection of articles dedicated to Alberto González Domínguez on his sixty-fifth birthday. Rev. Un. Mat. Argentina **25**, 137–143 (1970, 1971)
6. Coifman, R.R., Weiss, G.: Analyse harmonique non-commutative sur certains espaces homogènes. Étude de certaines intégrales singulières. Lecture Notes in Mathematics, vol. 242, Springer, Berlin (1971)
7. David, G., Journé, J.-L., Semmes, S.: Opérateurs de Calderón-Zygmund, fonctions para-accrétives et interpolation. Rev. Mat. Iberoamericana **1**(4), 1–56 (1985)
8. Folland, G.B.: A fundamental solution for a subelliptic operator. Bull. Amer. Math. Soc. **79**, 373–376 (1973)
9. Folland, G.B.: Subelliptic estimates and function spaces on nilpotent Lie groups. Ark. Mat. **13**(2), 161–207 (1975)
10. Folland, G.B.: On the Rothschild-Stein lifting theorem. Comm. Partial Differ. Equ. **2**(2), 165–191 (1977)
11. Folland, G.B.: Applications of analysis on nilpotent groups to partial differential equations. Bull. Amer. Math. Soc. **83**(5), 912–930 (1977)
12. Folland, G.B., Stein, E.M.: Estimates for the $\overline{\partial}_b$ complex and analysis on the Heisenberg group. Comm. Pure Appl. Math. **27**, 429–522 (1974)
13. Folland, G.B., Stein, E.M.: Parametrices and estimates for the $\overline{\partial}_b$ complex on strongly pseudo-convex boundaries. Bull. Amer. Math. Soc. **80**, 253–258 (1974)
14. Goodman R.W.: Lifting vector fields to nilpotent Lie groups. J. Math. Pures Appl. **57**(1, 9), 77–85 (1978)
15. Hörmander, L., Melin, A.: Free systems of vector fields. Ark. Mat. **16**(1), 83–88 (1978)
16. Kaplan, A.: Fundamental solutions for a class of hypoelliptic PDE generated by composition of quadratic forms. Trans. Amer. Math. Soc. **258**(1), 147–153 (1980)
17. Knapp, A.W., Stein, E.M.: Intertwining operators for semisimple groups. Ann. Math. **93**(2), 489–578 (1971)
18. Lanconelli, E., Polidoro, S.: On a class of hypoelliptic evolution operators. Partial differential equations, II (Turin, 1993). Rend. Sem. Mat. Univ. Politec. Torino **52**(1), 29–63 (1994)
19. Rothschild, L.P., Stein, E.M.: Hypoelliptic differential operators and nilpotent groups. Acta Math. **137**(3–4), 247–320 (1976)
20. Stein, E.M.: Harmonic analysis: Real-Variable Methods, Orthogonality, and Oscillatory Integrals. With the assistance of Timothy S. Murphy. Princeton Mathematical Series, 43. Monographs in Harmonic Analysis, III. Princeton University Press, Princeton (1993)
21. Stein, E.M.: Singular Integrals and Differentiability Properties of Functions. Princeton Mathematical Series, No. 30 Princeton University Press, Princeton (1970)

Chapter 4
Geometry of Hörmander's Vector Fields

As the paper [32] by Rothschild-Stein demonstrates, the study of general Hörmander's operators, in absence of an underlying structure of homogeneous group, can be very difficult. A general idea which has proved to be fruitful is that of exploiting the geometry of the operator as is written in the vector fields themselves: the properties of the integral lines of the vector fields and the metrics defined by them can be useful to describe sharp pointwise bounds on fundamental solutions or other local properties related to the operator, as well as to define function spaces suitable to express a control of the solutions. The study of geometry of Hörmander's vector fields, and particularly of the metrics induced by them, began flourishing with some deep papers of the middle 1980s, which we will briefly describe in Sect. 4.2, but also has some older antecedents in the study of integral lines of vector fields, the notion of connectivity, and the study of propagation of maxima. We will start (Sect. 4.1) explaining these antecedents, which also give us the opportunity of pointing out some connections of the geometry of Hörmander's vector fields with the applied sciences.

4.1 Connectivity, and Some of its Meanings

4.1.1 Exponential of a Vector Field, and How to Move Along the Direction of a Commutator

We now would like to give some insight of why Hörmander's condition can be related to the regularizing property of Hörmander's operators. Let us start with the following (standard) definition of *exponential of a vector field*.

Definition 54 *Let X be any vector field defined in a domain $\Omega \subset \mathbb{R}^n$ and, say, $C^1(\Omega)$. For $x_0 \in \Omega$, $t > 0$ let $\phi(s)$ be the solution to the Cauchy problem*

M. Bramanti, *An Invitation to Hypoelliptic Operators and Hörmander's Vector Fields*, SpringerBriefs in Mathematics, DOI: 10.1007/978-3-319-02087-7_4, © The Author(s) 2014

$$\begin{cases} \phi'(s) = X_{\phi(s)} \\ x(\phi) = x_0. \end{cases}$$

defined in $[0, t]$ *for t small enough. We set*

$$\exp(tX)(x_0) = \phi(t).$$

Equivalently: let x(s) be the solution to the Cauchy problem

$$\begin{cases} x'(s) = tX_{x(s)} \\ x(0) = x_0. \end{cases}$$

For t small enough the solution x(s) will be defined at least on the interval $s \in [0, 1]$. *We set*

$$\exp(tX)(x_0) = x(1).$$

Proof of the equivalence: for x as above, set $\phi(s) = x(s/t)$ for $s \in [0, t]$. Then

$$\phi'(s) = x'(s/t)/t = tX_{x(s/t)}/t = X_{\phi(s)}; \quad \phi(0) = x(0) = x_0,$$
$$\phi(t) = x(1) = \exp(tX)(x_0).$$

Analogously one proves the converse implication.

With the above notation, a standard result about ODEs reads as follows:

Theorem 55 *Let X, Y be a couple of* C^1 *vector fields in a domain* Ω. *Then*

$$\exp(-tX)\exp(-tY)\exp(tX)\exp(tY)(x_0) = \exp\left(t^2[Y, X]\right)(x_0) + o\left(t^2\right)$$
$$= x_0 + t^2[Y, X]_{x_0} + o\left(t^2\right)$$

as $t \to 0^+$.

This means that moving repeatedly along the integral curves of the vector fields $tY, tX, -tY, -tX$ causes as a net result a smaller displacement in the approximate direction of the commutator $t^2[Y, X]$. In this sense a control in the directions of a family of Hörmander's vector fields can give a control in *any* direction.

Example 56 *Let us check this property in the simplest case, namely that of Grushin vector fields:*

$$X = \partial_x; \quad Y = x\partial_y.$$

At the origin we have $X_{(0,0)} = \partial_x$, $Y_{(0,0)} = 0$, *hence following the integral curves of these vector fields it is impossible to move directly in the y direction. But let us move along the integral curves of X and Y as in the above theorem, that is let us compute:*

$$\exp(-tX)\exp(-tY)\exp(tX)\exp(tY)(0, 0).$$

It is more convenient here to write X, Y as:

$$X = (1, 0); \quad Y = (0, x).$$

Step 1. $\exp(tY)(0,0)$. *Let* $u = (x, y)$. *We have to solve*

$$\begin{cases} u'(s) = tY_{u(s)} \\ u(0) = (0,0) \end{cases} \text{ that is } \begin{cases} x'(s) = 0 \\ y'(s) = tx(s) \\ (x,y)(0) = (0,0). \end{cases}$$

The first equation gives

$$x(s) = const. = 0, \quad hence\ y' = 0, y = 0$$

and $\exp(tY)(0) = 0$.

Step 2. $\exp(tX)\exp(tY)(0) = \exp(tX)(0)$. *We have to solve*

$$\begin{cases} u'(s) = tX_{u(s)} \\ u(0) = (0,0) \end{cases} \text{ that is } \begin{cases} x'(s) = t \\ y'(s) = 0 \\ (x,y)(0) = (0,0). \end{cases}$$

hence $y(s) = const. = 0, x(s) = ts;$

$$\exp(tX)\exp(tY)(0) = (x(1), y(1)) = (t, 0).$$

Step 3. $\exp(-tY)\exp(tX)\exp(tY)(0,0) = \exp(-tY)(t,0)$. *We have to solve*

$$\begin{cases} u'(s) = -tY_{u(s)} \\ u(0) = (t,0) \end{cases} \text{ that is } \begin{cases} x'(s) = 0 \\ y'(s) = -tx \\ (x,y)(0) = (t,0). \end{cases}$$

hence $x(s) = const. = t, y'(s) = -t^2; y(s) = -t^2 s,$ *and*

$$\exp(-tY)\exp(tX)\exp(tY)(0,0) = (x(1), y(1)) = \left(t, -t^2\right).$$

Step 4. $\exp(-tX)\exp(-tY)\exp(tX)\exp(tY)(0,0) = \exp(-tX)\left(t, -t^2\right)$. *We have to solve*

$$\begin{cases} u'(s) = -tX_{u(s)} \\ u(0) = \left(t, -t^2\right) \end{cases} \text{ that is } \begin{cases} x'(s) = -t \\ y'(s) = 0 \\ (x,y)(0) = \left(t, -t^2\right). \end{cases}$$

hence $y(s) = const. = -t^2, x(s) = t - ts;$

$$\exp(-tX)\exp(-tY)\exp(tX)\exp(tY)(0,0) = (x(1), y(1)) = \left(0, -t^2\right).$$

Finally, since $[X, Y] = \partial_y = (0, 1)$, *we have*

$$\exp\left(t^2[Y, X]\right)(0, 0) = \left(0, -t^2\right)$$

which proves the assertion. In this case the remainder $o\left(t^2\right)$ *vanishes, and we have an exact equality.*

4.1.2 Rashevski-Chow's Connectivity Theorem

Once we has realized that a suitable movement along the integral curves of $tY, tX, -tY, -tX$ causes an actual displacement in the direction of $t^2[Y, X]$, it is quite natural to think that, composing in a suitable way movements along the integral curves of a system of Hörmander's vector fields one can obtain a displacement in the direction of any (also iterated) commutator, and therefore *in any direction*. This can be actually proved, arriving to the following well-known result:

Theorem 57 (Connectivity theorem) *Let* $X_1, X_2, \ldots, X_q$ *be any system of Hörmander's vector fields in an open connected set* $\Omega \subset \mathbb{R}^n$. *Then for every couple of points* $x, y \in \Omega$ *there exists an absolutely continuous curve* γ *contained in* Ω *and joining x to y, such that* γ *is composed by integral curves of the* X_i's.

The previous theorem is known as Rashevski-Chow theorem, from [8, 31] and is an old result, dating back to 1938-39. Although the study of Hörmander's operators did not exist yet at that time, these geometric properties of noncommuting vector fields had been studied in the context of Pfaffian systems, and had been motivated by the study of nonholonomic systems and by Carathéodory's work on the foundations of thermodynamics. This is a story which is worthwhile to be briefly retold.

4.1.3 Carathéodory Foundations of Thermodynamics and Inaccessibility

In 1909 Carathéodory wrote an essay on the mathematical foundations of thermodynamics. The physicist Max Born had urged him to do this, feeling that the current presentation of the matter was not logically and mathematically satisfactory.[1]

---

[1] The original paper by Carathéodory is [7], in German. An English translation can be found in the book [25, Chap. 12]. To the reader who is interested in Carathéodory's ideas on thermodynamics, however, I suggest the reading of some parts of an essay written by Max Born on that subject, see [4, Chap. V and Appendix 6, 7].

Carathéodory's work presents thermodynamics as an axiomatic theory.[2] The point we are interested in is the second axiom of the theory, which is designed to imply the second law of thermodynamics. Contenting ourselves with an informal account of these issues, we will skip all the previous definitions and come to the point. The second axiom, also known for its importance as *Carahtéodory principle*, states that:

If S is the state space of a thermodynamic system, for any $x_0 \in S$ and for any neighborhood $U(x_0)$ there exists a state $x \in U(x_0)$ such that x is not accessible from x_0 along an adiabatic path.

The author has previously discussed that an adiabatic path is any curve which is solution to the Pfaffian system[3]:

$$dQ = f_1(V_1, V_2, \vartheta) \, dV_1 + f_2(V_1, V_2, \vartheta) \, dV_1 + f_3(V_1, V_2, \vartheta) \, d\vartheta = 0$$

where V_i represents volumes of two portions of the system in thermal equilibrium and ϑ is the common temperature; hence in this case the space S is three-dimensional. Then Carathéodory proves a theorem about Pfaffian equations, stating that under Axiom 2 there exists an integrating factor of dQ, that is two scalar functions λ and ϕ such that

$$\lambda \, dQ = d\phi,$$

where $d\phi$ is an exact differential. Under suitable normalization, the function λ turns out to be interpretable as absolute temperature, while ϕ is the entropy. He has therefore proved the relation

$$dQ = \frac{dS}{T}$$

which is the core of the second law of thermodynamics. *Inaccessibility* (which is the opposite of connectivity) is therefore the differential geometric concept which contains the seed of the idea of *irreversibility* in thermodynamics.

Let us explore more the relation between these ideas and Hörmander's vector fields. To fix ideas, assume the system has $n + 1$ independent variables $x_0, x_1, \ldots, x_n$, so that

$$dQ = \sum_{i=0}^{n} A_i(x) \, dx_i.$$

Saying that a curve $x = \phi(t)$ is an adiabatic path means that

$$\sum_{i=0}^{n} A_i(\phi(t)) \, \phi_i'(t) \, dt = 0.$$

---

[2] Recall that one of Hilbert's problems posed in the international congress of mathematics held in Paris in 1900 consisted in giving an axiomatization of *physical* theories, following the general trend to axiomatization which pervaded mathematics in those years.

[3] Here we follow the simplified presentation of the theory given in [4, Chap. 5]; Carathéodory's approach is more general and abstract, and involves Pfaffian forms of n variables.

In other words, $\phi'(t)$ is normal to the vector field

$$(A_0, A_1, \ldots, A_n).$$

This can be rephrased saying that $\phi(t)$ is the integral line of a vector field which is linear combination of the n vector fields:

$$X_i = \partial_{x_i} - \frac{A_i}{A_0}\partial_{x_0} \text{ for } i = 1, 2, \ldots, n.$$

Actually,

$$(A_0, A_1, \ldots, A_n) \cdot \sum_{i=1}^{n} c_i X_i = (A_0, A_1, \ldots, A_n) \cdot \left(\sum_{i=1}^{n} \left(-\frac{c_i A_i}{A_0} \right), c_1, c_2, \ldots, c_n \right) = 0.$$

Now, if the n vector fields satisfied Hörmander's condition in $U(x_0)$, by Rashevski-Chow theorem any point $x \in U(x_0)$ would be accessible by x_0 along integral lines of the X_i's, that is along an adiabatic path. Since, by Carathéodory principle, we know that this is not the case, Hörmander's condition must fail in $U(x_0)$. To fix ideas, assume it fails *everywhere* in $U(x_0)$. Since we are considering n vector fields in a space of dimension $n + 1$, the failure of Hörmander's condition amounts to saying that any commutator of the X_i's is expressible as a linear combination of the X_i's themselves. This is also expressed saying that the system $\{X_i\}_{i=1}^{n}$ is *involutive*. A known result from differential geometry states that an involutive system is integrable, which is the same conclusion as Carathéodory theorem about Pfaffian forms, yielding the existence of absolute temperature and entropy. Namely, the $n + 1$ dimensional space S is foliated in n-hypersurfaces which contain integral lines of the vector fields. Since every open set contains infinitely many different leaves, it also contains points which are impossible to connect to each other.

Summarizing: in the particular case of a system of n vector fields in $n + 1$ dimensions, knowing that Hörmander's condition fails everywhere gives a "positive" piece of information: it says that the system of vector fields is involutive and hence integrable.

4.1.4 Connectivity, Controllability, and Nonholonomy

Here we will sketch the idea of controllability in the context of geometric control theory, to point out its connection with the concept of connectivity. Good references for geometric control theory can be found for instance in the books by Jurdjevic[4] [24], Bloch [2], Agrachev-Sachkov [1].

---

[4] For an overview of the subject I particularly suggest the reading of Jurdjevic' introduction to his book and the introduction to Chap. 1.

Let us consider a physical system whose dynamics is governed by a (nonlinear, autonomous) systems of ODEs:

$$x' = f(x)$$

where f is defined on a smooth n-dimensional manifold M, called the configuration space of the system. If we have some control on the system (like steer and pedals of our car), then the actual dynamics follows a more general law of the kind

$$x' = f(x, u) \tag{4.1}$$

where $u : [0, T] \to \mathbb{R}^m$ describes the controls and f is defined on $M \times \mathbb{R}^m$. In this case we can say that we have m controls on the system, which has n degrees of freedom. The general problem of control theory is to find a suitable control function $u : [0, T] \to \mathbb{R}^m$ (for some $T > 0$) which drives the system from some initial state x_0 to some desired final state x_f. This means that the solution to the Cauchy problem:

$$\begin{cases} x' = f(x, u(t)) \\ x(0) = x_0 \end{cases}$$

is defined at least on $[0, T]$ and $x(T) = x_f$.

We say that *the system is controllable* if for any couple of states $x_0, x_f \in M$ there exists a control function $u(t)$ which drives the system from x_0 to x_f in some time T.

Equation (4.1) is very general; one often studies systems which involve the controls in a simpler way, namely

$$x' = f(x) + \sum_{j=1}^{m} f_j(x) u_j \tag{4.2}$$

or

$$x' = \sum_{j=1}^{m} f_j(x) u_j. \tag{4.3}$$

The system (4.2) is called *nonlinear control affine system*; the term f is called *drift*, and describes some elements of the system which we cannot directly control.[5] The system (4.3) represents the particular case when the drift vanishes, and as we will see is directly related to Rashewski-Chow theorem. Assume that $u(t)$ is a control function that drives the system (4.3) from x_0 to x_f in time T; then the actual path of the system, in the configuration space, is a curve $x = \phi(t)$ which solves:

---

[5] A clear example of such control system is rowing a boat on a river: the river's drift is out of our control.

$$\begin{cases} \phi'(t) = \sum_{j=1}^{m} f_j\,(\phi\,(t))\,u_j\,(t) \\ x\,(0) = x_0 \end{cases}$$

and this means that ϕ is an integral curve of the family of vector fields $f_1, \ldots, f_m$.

Hence the control system coincides with a family of vector fields; the concept of controllability of the system (4.3) coincides with that of connectivity of M by means of integral curves of the vector fields $f_1, \ldots, f_m$. In particular, if the f_i's satisfy Hörmander's condition, then the system is controllable.

Example 58 (Car control) *(See [29, pp. 703–4]). Let us consider a car which can move in a plane. Let (x, y) be the middle point of the rear axle; l the distance between the two axles; ϑ the angle formed by the car with the x axis; ϕ the angle formed by the front wheels with respect to the car.*

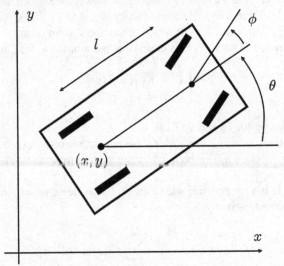

We can control the linear velocity u_1 of the car, and the the angle velocity of steering $\phi' = u_2$. Let us express the dynamics of the system. First of all, the point (x, y) moves according to

$$x' = u_1 \cos \vartheta$$
$$y' = u_1 \sin \vartheta.$$

We also know that

$$\phi' = u_2.$$

Note that we have not a direct control on the angle ϑ, but there is a nonholonomic constraint relating ϑ with ϕ, which allows us to write an equation for ϑ'. First, the quantity $l\vartheta'$ represents the linear velocity of the middle point of the front axle in the direction normal to the longitudinal axis of the car; the actual instantaneous velocity of this point is a vector $\mathbf{v}$ which forms an angle ϕ with the longitudinal axis of the *car; with respect to the car, its components are $l\vartheta'$ and u_1, hence*

$$l\vartheta' = u_1 \tan \phi.$$

We have therefore the control system

$$\begin{cases} x' = u_1 \cos \vartheta \\ y' = u_1 \sin \vartheta \\ \phi' = u_2 \\ \vartheta' = \frac{u_1}{l} \tan \phi \end{cases}$$

This can be rewritten, letting $\mathbf{r} = (x, y, \phi, \vartheta)$ as

$$\mathbf{r}' = u_1 X_1 + u_2 X_2$$

with

$$X_1 = \cos \vartheta \, \partial_x + \sin \vartheta \, \partial_y + \frac{1}{l} \tan \phi \partial_\vartheta$$

$$X_2 = \partial_\phi.$$

Let us check Hörmander's condition:

$$[X_2, X_1] = \frac{1}{l} \left[1 + \tan^2 \phi \right] \partial_\vartheta$$

$$[[X_2, X_1], X_1] = \frac{1}{l} \left[1 + \tan^2 \phi \right] \{ -\sin \vartheta \, \partial_x + \cos \vartheta \, \partial_y \}.$$

The four vectors $X_1, X_2, [X_2, X_1], [[X_2, X_1], X_1]$ are always independent, provided $\phi \neq \pm \frac{\pi}{2}$. Hence the system is controllable.

Remark 59 *The interesting feature of examples like the previous one consists in the fact that the system is controllable even though the number of controls at our disposal (in this case, 2) is less than the number of degrees of freedom of the system (in this case, 4). This fact is strictly related to the anholonomy of the constraint. A system with 4 degrees of freedom and one holonomic constraint would have 3 free variables, and would be impossible to control with less than three controls. In the terminology derived from geometric control theory, a system of vector fields satisfying Hörmander's condition is also called* completely nonholonomic.[6]

Just to point out that the general problem of controllability does not always reduces to that of connectivity with respect to a family of vector fields, let us turn for a moment to a general affine control system (4.2) (with a drift). In this case, an admissible path will be a curve $x = \phi(t)$ which solves:

---

[6] Actually, Gromov in [19, p. 87] suggests that some form of the connectivity theorem could be known to Lagrange in the context of nonholonomic mechanics.

$$\begin{cases} \phi'(t) = f(\phi(t)) + \sum_{j=1}^{m} f_j(\phi(t)) u_j(t) \\ x(0) = x_0 \end{cases}$$

and this means that ϕ is a *particular* integral curve of the family of vector fields $f, f_1, \ldots, f_m$, the peculiarity being that now the curve can never stop or go backwards with respect to f: the system is forced to evolve continuously along the positive direction of f. Hence the validity of Hörmander's condition is no longer a sufficient condition for controllability, but needs to be supplemented by other assumptions.

Example 60 (Control of a rigid body by means of jet torques) *(See [24, Chap. 4, §3, §6]). Let us consider a rigid body (think to a space satellite) which is free to move around its center of gravity. Euler's equations for the angular velocities ω_i ($i = 1, 2, 3$) read as follows:*

$$\begin{cases} \omega_1' = \frac{I_2 - I_3}{I_1} \omega_2 \omega_3 \\ \omega_2' = \frac{I_3 - I_1}{I_2} \omega_1 \omega_3 \\ \omega_3' = \frac{I_1 - I_2}{I_3} \omega_1 \omega_2 \end{cases}$$

where I_i ($i = 1, 2, 3$) are the principal momenta of inertia of the body. Assume we can control the angular velocity with respect of one, two or three of the principal axes by means of jet torques. It is quite obvious that if we can use three (couples of) jets we can completely control the system. An interesting problem is: can we still control the system with less than three jet torques, for instance with two of them? The control system is in this case:

$$\begin{cases} \omega_1' = \frac{I_2 - I_3}{I_1} \omega_2 \omega_3 + a_1 u_1 \\ \omega_2' = \frac{I_3 - I_1}{I_2} \omega_1 \omega_3 + a_2 u_2 \\ \omega_3' = \frac{I_1 - I_2}{I_3} \omega_1 \omega_2 \end{cases} \tag{4.4}$$

where a_1, a_2 are nonzero constants, and u_1, u_2 are the controls. We can rewrite (4.4) as an affine control system, as follows:

$$\omega' = u_1 X_1 + u_2 X_2 + X_0$$

where

$$X_1 = a_1 \partial_{\omega_1}; \ X_2 = a_2 \partial_{\omega_2};$$
$$X_0 = \frac{I_2 - I_3}{I_1} \omega_2 \omega_3 \partial_{\omega_1} + \frac{I_3 - I_1}{I_2} \omega_1 \omega_3 \partial_{\omega_2} + \frac{I_1 - I_2}{I_3} \omega_1 \omega_2 \partial_{\omega_3}.$$

Let us check Hörmander's condition:

$$[X_1, X_0] = a_1 \frac{I_3 - I_1}{I_2} \omega_3 \partial_{\omega_2} + a_1 \frac{I_1 - I_2}{I_3} \omega_2 \partial_{\omega_3};$$

$$[X_2, [X_1, X_0]] = a_1 a_2 \frac{I_1 - I_2}{I_3} \partial_{\omega_3}$$

hence $X_1, X_2, [X_2, [X_1, X_0]]$ span $\mathbb{R}^3$ provided $I_1 \neq I_2$. This means that if the three principal momenta of inertia are not all equal, it is possible to choose two jet torques such that Hörmander's condition is fulfilled. As we have already said, this condition alone is not sufficient to assure controllability for an affine system; however, this particular drift can be proved to have a further property called recurrence (see [24, Chap. 4, §6]), which coupled with Hörmander's condition actually implies controllability. Hence two jet torques are enough to control a rigid body which has at least two different principal momenta of inertia.

4.1.5 Propagation of Maxima

A feature of Hörmander's operators, which is related to the geometry of vector fields and enlightens the importance of Hörmander's condition, is the way Hörmander's operators *propagate maxima*. This fact has been studied by Bony [3], 1969. Consider an operator

$$L = \sum_{i=1}^{q} X_i^2 + X_0 + a$$

with $a(x) \leq 0$ in Ω, X_i, X_0 smooth vector fields (not a priori satisfying Hörmander's condition!). Then:

Theorem 61 (See [3, Thm. 3.1]) *Let $u \in C^2(\Omega)$ satisfy $Lu \geq 0$ in Ω. Let $Z \in \mathcal{L}(X_1, X_2, \ldots, X_q)$ (this symbol denoting the Lie algebra generated by these vector fields) and let Γ be an integral curve of Z. If the maximum of u in Ω is positive and is attained at one point of Γ, then it is attained at all points of Γ.*

Clearly, if the vector fields $X_1, X_2, \ldots, X_q$ satisfy Hörmander's condition in Ω, then any smooth curve in Ω is an integral curve of some $Z \in \mathcal{L}(X_1, X_2, \ldots, X_q)$, hence by the previous theorem the maximum propagates in all Ω, and L satisfies a strong maximum principle. But the theorem is interesting in itself since it shows how "information travels along the integral curves of the vector fields *and their commutators*". The above theorem does not explain the role of X_0; this is the task of the next one:

Theorem 62 (See [3, Thm. 3.2]) *Let $u \in C^2(\Omega)$ satisfy $Lu \geq 0$ in Ω. Let $Z \in \mathcal{L}(X_1, X_2, \ldots, X_q)$ and let $x(t)$ be a curve satisfying $x'(t) = Z(x(t)) + \lambda(t) X_0(x(t))$ for some smooth positive function λ. If the maximum of u in Ω is* *positive and is attained at some point $x(t_0)$, then it is attained at all points $x(t)$ for $t \geq t_0$.*

In this case if the vector fields $X_1, X_2, \ldots, X_q, X_0$ satisfy Hörmander's condition in Ω, then the operator L satisfies a strong maximum principle "of parabolic type": maxima propagates in all the directions accessible by elements of $\mathcal{L}(X_1, X_2, \ldots, X_q)$ but only *inward* along X_0.[7]

The comparison between the two theorems also gives a flavour of the important differences which can exist between operators which are "sum of squares" of Hörmander's vector fields (possibly with lower order terms) and operators where the drift X_0 is needed in order to fulfill Hörmander's condition.

4.2 Metric Balls Induced by Systems of Vector Fields

4.2.1 Motivation

We have seen that a left invariant 2-homogeneous Hörmander's operator on a homogeneous group,

$$L = \sum_{i=1}^{q} X_i^2 + X_0$$

possesses a $(2 - Q)$-homogeneous fundamental solution, which therefore has a singularity expressed by estimates like:

$$|\Gamma(x)| \leq \frac{c}{\|x\|^{Q-2}};$$

$$|X_i \Gamma(x)| \leq \frac{c}{\|x\|^{Q-1}}, \quad i = 1, 2, \ldots, q;$$

$$|X_i X_j \Gamma(x)| + |X_0 \Gamma(x)| \leq \frac{c}{\|x\|^{Q}}.$$

In absence of an underlying homogeneous group, we have seen that the operator L can be (locally) lifted to an operator

$$\widetilde{L} = \sum_{i=1}^{q} \widetilde{X}_i^2 + \widetilde{X}_0$$

which possesses a parametrix of the kind

---

[7] The reader will recognize some analogy with the controllability problem for an affine control system with drift, discussed in the previous paragraph.

$$P\left(\xi,\eta\right)=\Gamma\left(\Theta_{\eta}\left(\xi\right)\right)$$

where Γ is still the homogeneous fundamental solution of a Hörmander's operator on a suitable homogeneous group, and $\Theta_{\eta}\left(\cdot\right)$ is a local diffeomorphism, so that our parametrix, although not homogeneous, still satisfies analogous growth estimate like

$$\left|P\left(\xi,\eta\right)\right|\leq\frac{c}{\left\|\Theta_{\eta}\left(\xi\right)\right\|^{Q-2}}\equiv\frac{c}{\rho\left(\xi,\eta\right)^{Q-2}};$$

$$\left|\widetilde{X}_i P\left(\cdot,\eta\right)\left(\xi\right)\right|\leq\frac{c}{\rho\left(\xi,\eta\right)^{Q-1}},\quad i=1,2,\dots,q;$$

$$\left|\widetilde{X}_i\widetilde{X}_j P\left(\cdot,\eta\right)\left(\xi\right)\right|+\left|\widetilde{X}_0 P\left(\cdot,\eta\right)\left(\xi\right)\right|\leq\frac{c}{\rho\left(\xi,\eta\right)^{Q}}.$$

However, what can we say about the original (unlifted) Hörmander's operator? If $G\left(x,y\right)$ denotes its local fundamental solution, does G satisfy any local growth estimate analogous to the above ones? The point is that in this situation there is no analog of the exponent Q. Experience with the abstract study of singular integrals in spaces of homogeneous type teaches that, in absence of an exponent Q of "homogeneous dimension",

the right expression analogous to $\dfrac{1}{\rho\left(\xi,\eta\right)^{Q}}$ should be $\dfrac{1}{\left|B\left(x,d\left(x,y\right)\right)\right|}$, and

the right expression analogous to $\dfrac{1}{\rho\left(\xi,\eta\right)^{Q-k}}$ should be $\dfrac{d\left(x,y\right)^{k}}{\left|B\left(x,d\left(x,y\right)\right)\right|}$

for $k=1,2$, provided d is the "right" distance (or quasidistance), and B is the corresponding metric ball. Hence the idea arises that, in order to prove sharp upper bounds on the local fundamental solution and its derivatives with respect to the vector fields for general Hörmander's operators, one should first of all introduce a good "distance induced by the vector fields" and study the properties of this distance and the corresponding metric balls. This program has been carried out first in a couple of fundamental papers of the middle 1980s, namely:

1984, A. Sanchez-Calle: Fundamental solutions and geometry of sum of squares of vector fields. Inventiones Math. [35]

1985, A. Nagel, E. M. Stein, S. Wainger: Balls and metrics defined by vector fields I: Basic properties. Acta Math. [30]

which were soon followed by other deep results, contained in

1986, D. S. Jerison, The Poincaré inequality for vector fields satisfying Hörmander's condition. Duke Math. J. [22]

1986, C. Fefferman, A. Sánchez-Calle: Fundamental solutions for second order subelliptic operators. Ann. of Math. [10]

1986, D. S. Jerison, A. Sánchez-Calle: Estimates for the heat kernel for a sum of squares of vector fields. Indiana Univ. Math. J. [23].

4.2.2 *Distance Induced by a System of Hörmander's Vector Fields*

We start describing some of the ideas contained in the paper by Nagel, Stein and Wainger [30]. Let us begin fixing some notation which will be useful in the following. Let $X_1, X_2, \ldots, X_q$ be a system of vector fields satisfying Hörmander's condition at step r in some open connected Ω of $\mathbb{R}^n$. For any multiindex $I = (i_1, i_2, \ldots, i_k)$ of *length* $|I| = k$ we set:

$$X_I = X_{i_1} X_{i_2} \ldots X_{i_k}$$

and

$$X_{[I]} = \left[X_{i_1}, \left[X_{i_2}, \ldots \left[X_{i_{k-1}}, X_{i_k} \right] \ldots \right] \right].$$

If $I = (i_1)$, then

$$X_{[I]} = X_{i_1} = X_I.$$

As usual, $X_{[I]}$ can be seen either as a differential operator or as a vector field. We will write

$$X_{[I]} f$$

to denote the differential operator $X_{[I]}$ acting on a function f, and

$$\left(X_{[I]} \right)_x$$

to denote the vector field $X_{[I]}$ evaluated at the point x. By Hörmander's condition, the vectors

$$\left\{ \left(X_{[I]} \right)_x \right\}_{|I| \leq r}$$

span $\mathbb{R}^n$ for any $x \in \Omega$. Hence for any absolutely continuous curve $\varphi : [0, 1] \longrightarrow \Omega$ there exist measurable functions $\{a_I(t)\}_{|I| \leq r}$ defined in $[0, 1]$ such that

$$\varphi'(t) = \sum_{|I| \leq r} a_I(t) \left(X_{[I]} \right)_{\varphi(t)} \quad \text{a.e.} \, t \in [0, 1].$$

With this in mind, we can define the *subelliptic metric* introduced by Nagel, Stein and Wainger [30]:

Definition 63 *For any $\delta > 0$, let $C(\delta)$ be the class of absolutely continuous mappings $\varphi : [0, 1] \longrightarrow \Omega$ which satisfy*

$$\varphi'(t) = \sum_{|I| \leq r} a_I(t) \left(X_{[I]} \right)_{\varphi(t)} \quad a.e.$$

with $a_I : [0, 1] \to \mathbb{R}$ measurable functions,

$$|a_I(t)| \leq \delta^{|I|}.$$

Then define

$$d(x, y) = \inf \{\delta > 0 : \exists \varphi \in C(\delta) \ with \ \varphi(0) = x, \varphi(1) = y\}.$$

By the remark before the definition, it is clear that the function $d : \Omega \times \Omega \to \mathbb{R}$ is finite. Moreover, it is a distance. This follows from the fact that the union of two consecutive admissible curves can be reparametrized to give an admissible curve.

The idea behind the above definition is the following: to reach a point y starting from x we can follow *any* curve we want, but we move faster if we choose to follow integral lines of the basic vector fields $X_1, \ldots, X_q$ ("the highways") and we move slower and slower as we follow integral lines of commutators of the X_i's of higher and higher step ("minor roads"). The distance d measures the total *time* we spend to reach y from x. It is quite easy to prove the following:

Proposition 64 (Relation with the Euclidean distance) *There exist a positive constant c_1 depending on Ω and the X_i's and, for every $\Omega' \Subset \Omega$, a positive constant c_2 depending on Ω' and the X_i's, such that*

$$c_1 |x - y| \leq d(x, y) \leq c_2 |x - y|^{1/r} \quad for \ any \ x, y \in \Omega'. \tag{4.5}$$

In particular, this means that the distance d induces Euclidean topology but is *not* equivalent to the Euclidean metric.

4.2.3 Volume of Metric Balls

One of the main results contained in [30] is an estimate of the volume of metric balls. Let us describe this result, together with a bit of its background. In order to study the volume of metric balls and other metric properties, it is useful to exploit suitable coordinate systems. Let $\left\{ \left(X_{[I]} \right)_{x_0} \right\}_{I \in B}$ be a basis of $\mathbb{R}^n$ obtained choosing, at the point x_0, suitable commutators of the X_i's. Let us consider the map

$$\{u_I\}_{I \in B} \mapsto x = \exp \left(\sum_{I \in B} u_I X_{[I]} \right)(x_0) \tag{4.6}$$

defined from a neighborhood of the origin in the space $\mathbb{R}^n$ of the variables $\{u_I\}_{I \in B}$ to a neighborhood $U(x_0)$. The Jacobian of this map at $u = 0$ equals the matrix $\left\{ \left(X_{[I]} \right)_{x_0} \right\}_{I \in B}$, which is nonsingular since the vectors $\left(X_{[I]} \right)_{x_0}$ are a basis of $\mathbb{R}^n$; hence this map is a local diffeomorphism, which represents a neighborhood of x_0 by means of the canonical coordinates $\{u_I\}_{I \in B}$. This coordinate system depends on the choice B of the basis. Note that, by definition of the distance d,

$$|u_I| \le \delta^{|I|} \ \forall I \in B \Longrightarrow x \in B(x_0, \delta).$$

Hence, to study metric balls, an interesting object is the δ-box

$$\left\{ u \in \mathbb{R}^n : |u_I| \le \delta^{|I|} \forall I \in B \right\}$$

(which also depends on our choice of B). Now, under the exponential mapping (4.6) the volume of the image of this box should be comparable to

$$\delta^{|B|} \left| \det \left\{ (X_{[I]})_{x_0} \right\}_{I \in B} \right| \tag{4.7}$$

having set

$$|B| = \sum_{I \in B} |I|.$$

This suggests that, in order to find a sharp estimate of the volume of metric balls $B(x_0, \delta)$, one has to choose the set B of generators that maximizes the quantity (4.7). Note that this choice depends *both on the center and on the radius of the ball*, which makes very delicate the analysis carried out by Nagel-Stein-Wainger. One could think that a simpler choice of B could work, namely *minimize* $|B|$. However, doing so we would find a value of (4.7) which for a fixed radius δ can vary *discontinuously* with x_0, even though the vector fields are smooth!

Example 65 *Let us consider the Grushin vector fields:*

$$X_1 = \partial_x; X_2 = x\partial_y \text{ in } \mathbb{R}^2.$$

There are just two interesting choices of B:

$$B_1 = ((1), (2)) \text{ (i.e. } X_1, X_2), \text{ which gives}$$
$$|B_1| = 2 \text{ and } \left| \det \left\{ (X_{[I]}) \right\}_{I \in B_1} \right| = |x|;$$
$$B_2 = ((1), (1, 2)) \text{ (i.e. } X_1 \text{ and } [X_1, X_2]), \text{ which gives}$$
$$|B_2| = 3 \text{ and } \left| \det \left\{ (X_{[I]}) \right\}_{I \in B_2} \right| = 1.$$

The first choice is possible only at points $x \ne 0$. Hence, at any point $(0, y_0)$ we will have to choose B_2, getting

$$\delta^{|B_2|} \left| \det \left\{ (X_{[I]})_{(0, y_0)} \right\}_{I \in B_2} \right| = \delta^3$$

while as soon as we move to (x_0, y_0) with $x_0 \neq 0$, choosing B_1 (which has smaller length than B_2) we would get

$$\delta^{|B_1|} \left| \det \left\{ \left(X_{[I]} \right)_{(x_0, y_0)} \right\}_{I \in B_1} \right| = \delta^2 |x_0|$$

which gives a jump. Note, however, that if we choose B_1 or B_2 according to which maximize (4.7), we would still choose B_2 for all $x_0 \in (-\delta, \delta)$.

We are now ready to state the result proved by Nagel-Stein-Wainger.

Theorem 66 (Volume of metric balls) *For any $\Omega' \Subset \Omega$ there exist constants C_1, C_2, r_0 such that for any $x \in \Omega'$ and $\delta \leq r_0$ one has*

$$0 < C_1 \leq \frac{|B(x, \delta)|}{\Lambda(x, \delta)} \leq C_2$$

where

$$\Lambda(x, \delta) = \sum_B \left| \det \left\{ \left(X_{[I]} \right)_x \right\}_{I \in B} \right| \delta^{|B|}$$

and the sum is taken over all the possible n-tuples B of multiindices I with $|I| \leq r$.

Example 67 *Let us consider again the Grushin vector fields:*

$$X_1 = \partial_x; \quad X_2 = x \partial_y \ in \ \mathbb{R}^2.$$

By what remarked in the previous example, the above theorem states that

$$C_1 \left(\delta^3 + \delta^2 |x| \right) \leq |B((x, y), \delta)| \leq C_2 \left(\delta^3 + \delta^2 |x| \right)$$

with C_1, C_2 depending on an upper bound on δ and $\sqrt{x^2 + y^2}$. In particular, the balls of center $(0, y_0)$ have volume comparable to δ^3, while the balls of center (x_0, y_0) with large x_0 and small radius δ have volume comparable to δ^2.

The proof of the above theorem is very long and delicate. In particular, several steps consist in proving suitable *uniform bounds from below* on the quantities (4.7), for suitable choices of the set B of generators. These bounds can be seen as a quantitative expression of the validity of Hörmander's condition, or also as a struggle against the degenerate character of the operator $\sum X_i^2$.

A relevant consequence of the above theorem is the following:

Corollary 68 (Local doubling property) *For any $\Omega' \Subset \Omega$ there exist constants C, r_0 such that for any $x \in \Omega'$ and $\delta \leq r_0$ one has*

$$|B(x, 2\delta)| \leq C |B(x, \delta)|.$$

It is worthwhile to note what happens in the particular case of a system of vector fields which are free up to step r and satisfy Hörmander's condition at step r in Ω. In this case, if B is a choice of generators such that

$$\det\left\{\left(X_{[I]}\right)_x\right\}_{I \in B} \neq 0 \tag{4.8}$$

at some point $x \in \Omega$, then this will be true at *every* point of Ω. Moreover, for a fixed $\Omega' \Subset \Omega$ the function $\left|\det\left\{\left(X_{[I]}\right)_x\right\}_{I \in B}\right|$ will have a positive lower (and upper) bound, hence the function $\Lambda\,(x, \delta)$ will be equivalent to $c\delta^Q$ for some Q which is the smallest value of $|B|$ such that (4.8) holds. Therefore:

Corollary 69 (Volume of metric balls for free vector fields) *If the X_i's are free up to step r and satisfy Hörmander's condition at step r in Ω then there exists a positive integer Q and, for any $\Omega' \Subset \Omega$, there exist positive constants C_1, C_2, r_0 such that for any $x \in \Omega'$ and $\delta \leq r_0$ one has*

$$C_1 \delta^Q \leq |B\,(x, \delta)| \leq C_2 \delta^Q.$$

4.2.4 The Control Distance

Another important result contained in the paper [30] by Nagel Stein Wainger is the equivalence between two different distances induced by the vector fields.[8] Recall that Chow's theorem (see Theorem 57) shows that it is possible to join any two points of Ω using only integral lines of the vector fields X_i. This suggests that one could also define a distance analogous to the d introduced above, but defined using only the basic vector fields $X_1, X_2, \ldots, X_q$. This justifies the following:

Definition 70 (Control distance, Carnot-Carathéodory distance) *For any $\delta > 0$, let $C_1\,(\delta)$ be the class of absolutely continuous mappings $\varphi : [0, 1] \longrightarrow \Omega$ which satisfy*

$$\varphi'\,(t) = \sum_{i=1}^{q} a_i\,(t)\,(X_i)_{\varphi(t)} \quad a.e.$$

with

$$|a_i\,(t)| \leq \delta \quad for\ i = 1, 2, \ldots, q.$$

We define

$$d_1\,(x, y) = \inf\left\{\delta > 0 : \exists \varphi \in C_1\,(\delta)\ with\ \varphi\,(0) = x, \varphi\,(1) = y\right\}.$$

---

[8] Actually the authors prove the equivalence between five different distances, but here we will concentrate just on two of them.

Compared with the previous definition of d, we see that now we can move only through the integral lines of the basic vector fields $X_1, X_2, \ldots, X_q$. Continuing the previous analogy, now all the streets are equally fast, but there are fewer streets, and it is not clear if we can get to any desired place.

Remark 71 *Note that the definition of d_1 does not require, a priori, the validity of Hörmander's condition, nor a high degree of regularity of the vector fields: if $\{X_i\}_{i=1}^q$ is any system of locally Lipschitz vector fields, then the definition of d_1 is meaningful. The function d_1 can be proved to be a distance, possibly infinite for some pairs of points. This distance is known in the literature as the* control distance *or the* Carnot-Carathéodory distance *induced by the system of vector fields $\{X_i\}_{i=1}^q$ and, as we will explain later (see Sect. 4.5), has become very important, through the years, for the development of abstract, axiomatic theories of systems of vector fields.*

If $\{X_i\}_{i=1}^q$ is actually a system of Hörmander's vector fields, then the distance d_1 is finite for any pair of points. This follows from the next

Theorem 72 (Local equivalence of distances) *The distances d, d_1 are locally equivalent, that is every $x_0 \in \Omega$ has a neighborhood $U(x_0)$ such that d and d_1 are equivalent in $U(x_0)$.*

In particular, one also has, locally,

$$c_1 |x - y| \leq d_1(x, y) \leq c_2 |x - y|^{1/r}$$

so that d_1 is still topologically, but not metrically, equivalent to the Euclidean distance. Also, the local doubling condition still holds for d_1-balls.

Note that the above theorem contains a quantitative version of the connectivity property expressed by Chow's theorem. This is well exemplified by the next interesting consequence, namely a version of "Lagrange theorem" with respect to the gradient $(X_1, \ldots, X_q)$:

Theorem 73 *Let $f \in C^1(B_1(x_0, \rho))$, then for any $x \in B_1(x_0, \rho)$ we have*

$$|f(x) - f(x_0)| \leq \rho \cdot \sup_{B_1(x_0, \rho)} |Xf|,$$

with

$$|Xf| = \sum_{i=1}^q |X_i f|.$$

In particular, this implies, for some constant $c > 0$,

$$|f(x) - f(x_0)| \leq cd(x, x_0) \cdot \sup_{B(x_0, cd(x_0, x))} |Xf|.$$

Proof. *Let $x \in B_1(x_0, \rho)$, then there exists a curve $\gamma(t)$ such that*

$$\gamma(0) = x_0; \gamma(1) = x$$

$$\gamma'(t) = \sum_{i=1}^{n} a_i(t) (X_i)_{\gamma(\tau)}$$

with $|a_i(t)| \leq \rho$, then

$$f(x) - f(x_0) = \int_0^1 \frac{d}{dt} (f(\gamma(t))) \, dt$$

$$= \int_0^1 \sum_{i=1}^{q} a_i(t) (X_i)_{\gamma(t)} \cdot \nabla f(\gamma(t)) \, dt$$

$$= \int_0^1 \sum_{i=1}^{q} a_i(t) (X_i f)(\gamma(t)) \, dt$$

$$|f(x) - f(x_0)| \leq \rho \int_0^1 \sum_{i=1}^{q} |(X_i f)(\gamma(t))| \, dt$$

$$\leq \rho \sup_{B_1(x_0,\rho)} \sum_{i=1}^{q} |X_i f|$$

$$\equiv \rho \sup_{B_1(x_0,\rho)} |X f| .$$

∎

Remark 74 (Carathéodory versus Carnot). *It has become customary to refer to the distance induced by a family of vector fields as to the Carnot-Carathéodory distance.*[9] *Now, while the link between Carathéodory's ideas on thermodynamics and the idea of connecting points moving along integral curves of vector fields is evident (as explained in Sect. 4.1.3), Carnot's pioneering works on thermodynamics rely on conceptions which are very far from these, as Born explains well in the following passage:*

"*Pfaffian equations are the mathematical expression of elementary thermal experiences, and one would expect that the laws of thermodynamics are connected with their properties. That is indeed the case, as Carathéodory has shown. But classical thermodynamics proceeded in quite a different way, introducing the conception of idealized thermal machines which transform heat into work and vice versa (William Thomson–Lord Kelvin), or which pump heat from a reservoir into another (Clau-*

---

[9] The first use of this term appeared in Gromov' book of 1981 "Structures métriques pour les variétés riemanniennes" (edited by J. Lafontaine and P. Pansu), later transformed in [20].

sius). The second law of thermodynamics is then derived from the assumption that not all processes of this kind are possible: you cannot transform heat completely into work, nor bring it from a state of lower temperature to one of higher 'without compensation'. These are new and strange conceptions, obviously borrowed from engineering." (See [4, p. 38]).

The only element that I have found in the literature supporting a relation with Carnot's ideas is the following passage by Gromov:

"The result [connectivity theorem] appears in 1909-paper by Carathéodory on formalization of the classical thermodynamics where horizontal curves roughly correspond to adiabatic processes. In fact the above proof [of the connectivity theorem] may be performed in the language of Carnot (cycles) and for this reason the metrics were christened "Carnot-Carathéodory" in [20]" (See [19, pp. 86–87]).[10]

4.2.5 Relation Between Lifted and Unlifted Balls

Let's go on with our account of the paper by Nagel-Stein-Wainger. The last part of the paper applies the previous results to the context studied by Rothschild-Stein in [32]. Let $X_1, X_2, \ldots, X_q$ be any system of Hörmander's vector fields in some neighborhood U of $x_0 \in \mathbb{R}^n$; we can consider the distance $d(x, y)$ induced in U by the X_i's; we will call it d_X now, to distinguish from other distances. Let $\widetilde{X}_1, \widetilde{X}_2, \ldots, \widetilde{X}_q$ be the free lifted vector fields defined as in [32] in a domain $\widetilde{U} = U \times I \subset \mathbb{R}^{n+m} = \mathbb{R}^N$. We can consider the distance $d_{\widetilde{X}}$ defined by the vector fields $\widetilde{X}_i$ in $\widetilde{U}$. Also, we know that in $\widetilde{U}$ there is a quasidistance $\rho(\xi, \eta) = \left\| \Theta_\eta(\xi) \right\|$ naturally attached to the $\widetilde{X}_i$'s, so two questions arise: which relation does exist between d_X and $d_{\widetilde{X}}$ and between ρ and $d_{\widetilde{X}}$?

Nagel-Stein-Wainger prove that:

Proposition 75 *The distance $d_{\widetilde{X}}$ and the quasidistance ρ, both defined in subsets of $\mathbb{R}^N$, are locally equivalent.*

More subtle is the relation between $d_{\widetilde{X}}$ and d_X, which are defined in subsets of spaces of different dimensions. First, they prove that

Proposition 76 *The projection of the metric ball $B_{\widetilde{X}}((x, t), r)$ on $\mathbb{R}^n$ is $B_X(x, r)$, which means that*

$$d_{\widetilde{X}}((x, t), (y, s)) \geq d_X(x, y).$$

It is not possible, however, to prove a control in the reverse sense, at the level of distances. For instance, a "reasonable" inequality like

$$d_{\widetilde{X}}((x, 0), (y, 0)) \leq c d_X(x, y)$$

---

[10] Gromov refers to the 1981 edition of [20], see the previous footnote.

has never been proved. What one can prove is a control (in both senses) at the level of *volumes*. This is a deep result, relying on the previous analysis of the structure of metric balls, and reads as follows:

Theorem 77 (Volumes of lifted and unlifted balls) *There exists $c_1 > 0$ such that for any $r > 0$ small enough and any $(x, t) \in \mathbb{R}^N$, $y \in \mathbb{R}^n$ (in the small neighborhoods under consideration),*

$$\left| B_{\widetilde{X}}((x,t),r) \right| \geq c_1 \left| B_X(x,r) \right| \left| \left\{ s \in \mathbb{R}^m : (y,s) \in B_{\widetilde{X}}((x,t),r) \right\} \right|.$$

Conversely, there exists $\delta \in (0, 1)$ and $c_2 > 0$ such that, for any r, (x, t) as above and $y \in B_X(x, \delta r)$,

$$\left| B_{\widetilde{X}}((x,t),r) \right| \leq c_2 \left| B_X(x,r) \right| \left| \left\{ s \in \mathbb{R}^m : (y,s) \in B_{\widetilde{X}}((x,t),r) \right\} \right|.$$

In the above statements, the symbol $|\cdot|$ stands for the full dimensional Lebesgue measure in any of the three spaces $\mathbb{R}^N$, $\mathbb{R}^n$, $\mathbb{R}^m$.

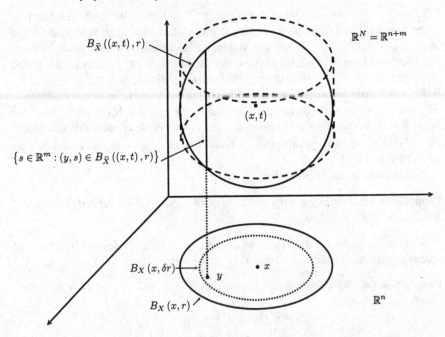

The geometric meaning of the theorem is that the volume of the lifted ball $B_{\widetilde{X}}$ is equivalent to the volume of a "cylinder" having the ball B_X as basis and the set $\left\{ s \in \mathbb{R}^m : (y,s) \in B_{\widetilde{X}} \right\}$ as "height". However, this equivalence of volumes is not the consequence of a simple set inclusion, but more the substitute of such lacking inclusion.

The above theorem is a powerful tool, which helps in deriving estimates in the original space starting from analogous estimates in the lifted space, where they are more easily established.

4.2.6 Estimates on the Fundamental Solution

The final result proved by Nagel-Stein-Wainger is a direct application of Theorem 77. Assume we have a kernel $k(\xi, \eta)$ in the lifted space, satisfying the local bound

$$|k((x,t),(y,s))| \leq c \frac{d_{\widetilde{X}}((x,t),(y,s))^a}{\left|B_{\widetilde{X}}\left((x,t),d_{\widetilde{X}}((x,t),(y,s))\right)\right|}$$

and let us define the "restricted kernel"

$$Rk(x,y) = \int_{\mathbb{R}^m} k((x,0),(y,s))\,\phi(s)\,ds$$

for a suitable cutoff function ϕ. Then it is not difficult to prove, using Theorem 77 and the doubling condition, that

$$|Rk(x,y)| \leq c \frac{d_X(x,y)^a}{\left|B_{\widetilde{X}}(x,d_X(x,y))\right|}.$$

Also, if

$$\left|\widetilde{X}_{i_1}\widetilde{X}_{i_2}\ldots\widetilde{X}_{i_h}k((x,t),(y,s))\right| \leq c \frac{d_{\widetilde{X}}((x,t),(y,s))^{a-h}}{\left|B_{\widetilde{X}}\left((x,t),d_{\widetilde{X}}((x,t),(y,s))\right)\right|}$$

then

$$\left|X_{i_1}X_{i_2}\ldots X_{i_h}Rk(x,y)\right| \leq c \frac{d_X(x,y)^{a-h}}{\left|B_{\widetilde{X}}(x,d_X(x,y))\right|}.$$

The previous general fact applies in particular to the parametrix built in [32] for the lifted operator, giving a local bound on a parametrix, and hence a fundamental solution, of the original operator.

This ends our account on the paper by Nagel-Stein-Wainger. Let us mention that the paper by Sanchez-Calle [35] has the same final goal (local estimates on the fundamental solution) and also contains Theorem 77, which however is proved by a completely different method. Moreover, in this paper the other properties of distances and metric balls are not proved.

Let us also mention the paper by Fefferman and Sánchez-Calle [10], which contains a far-reaching extension of the aforementioned upper bounds on fundamental solutions, for general subelliptic operators with nonnegative characteristic form (not necessarily written as sum of squares of vector fields).

4.3 Heat Kernels and Gaussian Estimates

We have already noted that if $L = \sum_{i=1}^{q} X_i^2$ is a Hörmander's operator in (a domain of) $\mathbb{R}^n$ then

$$H = \partial_t - \sum_{i=1}^{q} X_i^2$$

is a Hörmander's operator in (a domain of) $\mathbb{R}^{n+1}$. For the heat kernel $\mathcal{H}$ of H, the upper bounds proved in [30] or [35] read as follows:

$$\mathcal{H}(x, y) \le \frac{c}{\left| B_{\widetilde{X}}(x, d_X(x, y)) \right|};$$

$$\left| X_{i_1} X_{i_2} \dots X_{i_h} \partial_t^l \mathcal{H}(x, y) \right| \le c \frac{d_X(x, y)^{-h-2l}}{\left| B_{\widetilde{X}}(x, d_X(x, y)) \right|}.$$

Although these estimates express the right homogeneity we can expect from a heat kernel, they do not say anything about possible Gaussian-type estimates for $\mathcal{H}$. Results of this kind require an ad-hoc analysis.[11]

For heat operators of the kind

$$H = \partial_t - \sum_{i=1}^{q} X_i^2 \tag{4.9}$$

with X_i left invariant homogeneous vector fields on a Carnot group in $\mathbb{R}^n$, Gaussian bounds have been proved by Varopoulos ([39, 38], see also [37]):

$$\frac{1}{ct^{Q/2}} e^{-c\|y^{-1} \circ x\|^2/t} \le h(t, x, y) \le \frac{c}{t^{Q/2}} e^{-\|y^{-1} \circ x\|^2/ct} \tag{4.10}$$

for any $x, y \in \mathbb{R}^n, t > 0$, where Q is the homogeneous dimension of the group, and $\|\cdot\|$ any homogeneous norm of the group. Two-sided Gaussian estimates and a scaling invariant Harnack inequality for the operator

$$H = \partial_t - \sum_{i,j=1}^{q} X_i (a_{ij} X_j)$$

have been proved by Saloff-Coste and Stroock in [34], where $\{a_{ij}\}$ is a uniformly positive matrix with measurable entries, and the vector fields X_i are left invariant with respect to a connected unimodular Lie group with polynomial growth. In absence of a group structure, Gaussian bounds for operators (4.9) have been proved, on a compact manifold and for finite time, by Jerison and Sanchez-Calle [23], with an analytic approach (see also the previous

---

[11] The following bibliographical account is quoted from [5, p. 3].

partial result in [35]), and, on the whole $\mathbb{R}^{n+1}$, by Kusuoka-Stroock [26, 27], using the Malliavin stochastic calculus.

These estimates look as follows:

$$\frac{e^{-cd(x,y)/\sqrt{t-s}}}{c\,|B\,(x,\sqrt{t-s})|} \leq h\,(t,x;s,y) \leq \frac{ce^{-d(x,y)/c\sqrt{t-s}}}{|B\,(x,\sqrt{t-s})|} \quad \text{for} \quad t > s.$$

An extension of the previous Gaussian bounds to the class of nonvariational heat-type operators

$$H = \partial_t - \sum_{i,j=1}^{q} a_{ij} X_i X_j$$

is proved in the monograph [5], which we will comment in Chap. 5.

4.4 Poincaré's Inequality, and Some of its Consequences

In 1986, Jerison [22] proved that for a general family of Hörmander's vector fields a Poincaré's inequality holds. The precise form of this result is the following:

Theorem 78 *Let $X_1, X_2, \ldots, X_q$ be Hörmander's vector fields in a domain Ω of $\mathbb{R}^n$. Then for any $\Omega' \Subset \Omega$, $1 \leq p < \infty$ there exist positive constants C, r_0 such that for any $x_0 \in \Omega'$, $r \leq r_0$ and $B_r = B(x_0, r)$, any $u \in Lip\,(B)$ we have*

$$\left(\frac{1}{|B_r|} \int_{B_r} |u - u_{B_r}|^p \, dx\right)^{1/p} \leq Cr \left(\frac{1}{|B_r|} \int_{B_r} \left(\sum_{i=1}^{q} |X_i u|^2\right)^{p/2} dx\right)^{1/p}.$$

Here the ball B_r is meant with respect to the Carnot-Carathéodory distance (the distance d_1 of Nagel-Stein-Wainger)

Jerison's proof of this inequality (see [22, Thm. 2.1] for the case $p = 2$ and [22, §6] for $p \neq 2$) relies on the results by Rothschild and Stein [32] and Nagel, Stein and Wainger [30]; namely, he first establishes the result on homogeneous groups, then reduces the general case to this one by means of Rothschild-Stein's lifting and approximation technique; but in doing so, he also exploits some results contained in [30], and a careful inspection of some arguments contained in the same paper, to check the exact dependence of some constants on suitable parameters.

So far, we have described a set of "geometric results" established for Hörmander's vector fields. Let us summarize some of them: the Carnot-Carathéodory distance is finite (Chow's theorem), and is topologically equivalent to the Euclidean distance (see Eq. 4.5); for metric balls a local doubling condition holds (Nagel-Stein-Wainger); a Poincaré inequality holds with respect to the X-gradient (Jerison).

Experience with the theory of elliptic equations and Sobolev spaces has shown that the joint validity of these properties has important consequences. Starting with the late 1980s, a lot of research has been done about the "axiomatic" implications of these properties. We refer to Hajlasz-Koskela's monograph [21] for a good exposition and a rich source of further references on this area of research. Some of these results have been established in great generality, as axiomatic theories. For instance, it is well-known that, roughly speaking, the validity of the doubling condition and a Poincaré's inequality imply a Sobolev embedding. This fact has been proved, at different levels of generality, by Saloff-Coste [33], Garofalo and Nhieu [18], Franchi, Lu and Wheeden [17], Hajlasz and Koskela [21]. In turn, the doubling condition, Poincaré and Sobolev inequalities allow to reply the so-called Moser's iteration technique, and prove a Harnack inequality and a Hölder continuity result for local solutions to (elliptic or subelliptic) variational second order equations. An example of this kind of result in our setting is the following.

Let us consider a linear second order variational operator of the kind

$$Lu \equiv \sum_{i,j=1}^{q} X_i^* \left(a_{ij}(x) X_j u \right) \tag{4.11}$$

where $X_1, \ldots, X_q$ is a set of Hörmander's vector fields, X_i^* denotes the transposed operator of X_i, and $\{a_{ij}\}_{i,j=1}^{q}$ is a symmetric uniformly positive definite matrix of $L^\infty(\Omega)$ functions:

$$\lambda |\xi|^2 \leq \sum_{i,j=1}^{q} a_{ij}(x) \xi_i \xi_j \leq \lambda^{-1} |\xi|^2$$

for some $\lambda > 0$, any $\xi \in \mathbb{R}^q$, a.e. $x \in \Omega$. We say that u is a local solution to the equation $Lu = 0$ in Ω if

$$u \in W_{X,loc}^{1,2}(\Omega) = \left\{ u \in L_{loc}^2(\Omega) : X_i u \in L_{loc}^2(\Omega) \text{ for } i = 1, 2, \ldots, q \right\}$$

and

$$\int_\Omega \sum_{i,j=1}^{q} a_{ij} X_i u X_j \varphi \, dx = 0 \text{ for any } \varphi \in C_0^\infty(\Omega).$$

Then:

Theorem 79 *Let u be a local solution to $Lu = 0$ in Ω. Then:*

(i) u is locally bounded, with

$$\|u\|_{L^\infty(B)} \leq c \left(\frac{1}{|2B|} \int_{2B} |u(x)|^2 \, dx \right)^{1/2}$$

for any $2B \subset \Omega$, with c depending on the coefficients a_{ij} only through the number λ.

(ii) If u is positive in Ω, then it satisfies a Harnack's inequality:

$$\sup_B u \leq c \inf_B u$$

for any $2B \subset \Omega$, with c depending on the coefficients a_{ij} only through the number λ.

(iii) u is Hölder continuous (in the usual, Euclidean sense) of some exponent $\alpha \in (0, 1)$, on any subset $\Omega' \Subset \Omega$:

$$|u(x) - u(y)| \leq c |x - y|^\alpha$$

for any $x, y \in \Omega'$, with c, α depending on Ω' and depending on the coefficients a_{ij} only through the number λ, and c also depending on $\|u\|_{L^2(\Omega)}$.

4.5 Carnot-Carathéodory Spaces

As anticipated above, starting with the 1980s the study of geometry of vector fields has taken also an axiomatic approach: one can consider a family of locally Lipschitz continuous vector fields in a domain of $\mathbb{R}^n$, and the Carnot-Carathéodory distance (briefly, CC-distance) induced by them, which a priori could be infinite, but which is usually assumed (axiomatically) to be finite for any pair of points. This setting is, basically, what is called a Carnot-Carathéodory space. Other natural assumptions which can be done are the following:

- assuming that the CC-distance is continuous with respect to the Euclidean distance (or, stronger assumption, assuming that the two distances are topologically equivalent);
- assuming that the space is endowed with a Borel measure which is doubling with respect to CC-balls (or, stronger assumption, assuming that this measure is the Lebesgue measure);
- assuming that some form of Poincaré inequality holds, with respect to the vector fields;
- and so on.

In this setting one can define: function spaces induced by the vector fields (Sobolev-type space) or by their metric (Hölder or Lipschitz spaces) or by their metric and the measure (BMO-type spaces); perimeters of sets and BV functions; Hausdorff dimension; geodesics; and a wealth of other notions and properties from real analysis, geometric measure theory, differential geometry, and so on. Then one can study the relations between these concepts and properties under some reasonable set of assumptions. Possible applications of these theories to second order differential equations structured on vector fields usually deal with operators in divergence form, like (4.11).

The literature of this kind is really broad by now, and it is impossible here to summarize its themes and results. A few references, which are both interesting in themselves and a good starting point for further bibliographic research, are: the already quoted monograph [21] by Hajlasz and Koskela, the survey [11] by Franchi, the paper [18] by Garofalo and Nhieu, the more recent monograph [36], by Sawyer and Wheeden, just to quote a few. The area of this field of research which is more related to differential geometry is also known as *subriemannian geometry*. Here authoritative references are those by Gromov, see [19, 20]; more recent books on this subject are [28], by Montgomery, and [6], by Calin and Chang.

Since the main theme of these lectures is the study of second order PDEs (built with or related to Hörmander's vector fields) we will not pursue further the discussion of "abstract" Carnot-Carathéodory spaces. Instead, a question we want to address is: what are the concrete situations where, for a given family of vector fields, we know that a reasonably rich set of assumptions of the types quoted above is actually satisfied? Clearly Hörmander's vector fields are one of these situations, but what else?

4.6 Franchi–Lanconelli Operators with Diagonal Vector Fields

Let us consider the following degenerate elliptic operator in two variables:

$$L = \partial_{x_1}^2 + |x_1|^{2\alpha} \partial_{x_2}^2 \quad \text{for some } \alpha > 0.$$

This can be seen as a "sum of squares of vector fields"

$$X_1 = \partial_{x_1}; X_2 = |x_1|^{\alpha} \partial_{x_2}.$$

Note, however, that as soon as α is not an even integer, these vector fields are not smooth. Anyhow, one can still define the CC distance induced by them, and it is easy to check that the connectivity property holds: the only obstacle to the free movement in the plane, following the integral lines of X_1, X_2 is due to the zero velocity of X_2 on the line $x_1 = 0$; but we can always come out from this line following X_1, so we can actually go from any P_1 to any P_2 along integral lines of X_1, X_2. An analogous phenomena appears with the more general operator

$$L = \Delta_{x_1} + |x_1|^{2\alpha} \Delta_{x_2} \text{ in } \mathbb{R}^{n_1 + n_2} \ni (x_1, x_2).$$

These are just simple examples of operators belonging to some classes of degenerate elliptic operators which have been studied by Franchi-Lanconelli in a series of 5 papers appeared in the years 1982 to 1985 (see [12–16]). Here we will not state precisely their assumptions and results, but just give some ideas. Basically, the authors consider a family of n vector fields in $\mathbb{R}^n$ of *diagonal form*, that is

$$X_i = \lambda_i(x) \partial_{x_i}\ i = 1, 2, \ldots, n \tag{4.12}$$

where the functions λ_i satisfy a suitable set of assumptions which anyhow allow them to vanish somewhere and to be nonsmooth. The authors introduce the distance induced by these vector fields, study the connectivity property, establish a doubling property, prove suitable Sobolev embedding, a Poincaré's inequality, and deduce a kind of De Giorgi-Nash-Moser theory for divergence operators structured on these vector fields. Note that these papers appeared before Nagel-Stein-Wainger's paper [30], so they can be considered pioneering works about the notion of *metric induced by a vector field*. Moreover, they actually consider nonsmooth vector fields, hence these results cannot be deduced by those established later for general Hörmander's vector fields. Also, these papers cannot be framed in the theory of weighted degenerate elliptic operators

$$Lu = \left(a_{ij}(x) u_{x_i}\right)_{x_j}$$

with

$$\lambda \omega(x) |\xi|^2 \le a_{ij}(x) \xi_i \xi_j \le \Lambda \omega(x) |\xi|^2,$$

studied e.g. by Fabes, Kenig and Serapioni [9] under the assumption that ω be an A_2-weight of Muckenhoupt, that is:

$$\left(\frac{1}{|B_r|} \int_{B_r} \omega(x)\, dx\right) \left(\frac{1}{|B_r|} \int_{B_r} \frac{dx}{\omega(x)}\right) \le C$$

independently of the ball B_r. The reason is that typical functions $\lambda_i(x)$ appearing in (4.12) are monomials, for which $1/\lambda_i$ is not locally integrable. On the other hand, the vector fields considered by Franchi–Lanconelli have a very particular structure, which makes hard to generalize their techniques to broader classes of operators. We will come back later on these papers when dealing with the recent theory of nonsmooth Hörmander's vector fields, in Chap. 5.

References

1. Agrachev, A.A., Sachkov, Y.L.: Control theory from the geometric viewpoint. Encyclopaedia of Mathematical Sciences, vol. 87. Control Theory and Optimization, II. Springerg, Berlin (2004)
2. Bloch, A.M.: Nonholonomic mechanics and control. With the collaboration of J. Baillieul, P. Crouch and J. Marsden. With scientific input from P. S. Krishnaprasad, R. M. Murray and D. Zenkov. Interdisciplinary Applied Mathematics, vol. 24. Systems and Control. Springer, New York (2003)
3. Bony, J.M.: Principe du maximum, inégalite de Harnack et unicité du problème de Cauchy pour les opérateurs elliptiques dégénérés. Ann. Inst. Fourier (Grenoble) 19 1969 fasc. 1, 277–304 xii
4. Born, M.: Natural Philosophy of Cause and Chance. Clarendon Press, Oxford (1948)

5. Bramanti, M., Brandolini, L., Lanconelli, E., Uguzzoni, F.: Non-divergence equations structured on Hörmander vector fields: heat kernels and Harnack inequalities. Mem. AMS **204**(961), 1–136 (2010)
6. Calin, O., Chang, D.-C.: Sub-Riemannian geometry. General theory and examples. Encyclopedia of Mathematics and its Applications, vol. 126. Cambridge University Press, Cambridge (2009)
7. Carathéodory, C.: Untersuchungen über die Grundlagen der Thermodynamik. Math. Ann. **67**(3), 355–386 (1909). An English translation can be found in the book [25, Chap. 12]
8. Chow, W.-L.: Über Systeme von linearen partiellen Differentialgleichungen erster Ordnung. Math. Ann. **117**, 98–105 (1939)
9. Fabes, E.B., Kenig, C.E., Serapioni, R.P.: The local regularity of solutions of degenerate elliptic equations. Comm. Partial Differ. Equ. **7**(1), 77–116 (1982)
10. Fefferman, C., Sánchez-Calle, A.: Fundamental solutions for second order subelliptic operators. Ann. Math. (2) **124**(2), 247–272 (1986)
11. Franchi, B.: BV spaces and rectifiability for Carnot-Carathéodory metrics: an introduction. Notes for a course at the Spring School on Nonlinear Analysis, Function Spaces and Applications 7, Prague, July 17–22, 2002. Downloadable at: http://www.mate.polimi.it/scuolaestiva/bibliografia/franchi_corso_NAFSA7.pdf
12. Franchi, B., Lanconelli, E.: De Giorgi's theorem for a class of strongly degenerate elliptic equations. Atti Accad. Naz. Lincei Rend. Cl. Sci. Fis. Mat. Natur. (8) **72** (1982), no. 5, 273–277 (1983)
13. Franchi, B., Lanconelli, E.: Hölder regularity theorem for a class of linear nonuniformly elliptic operators with measurable coefficients. Ann. Scuola Norm. Sup. Pisa Cl. Sci. (4) **10**(4), 523–541 (1983)
14. Franchi, B., Lanconelli, E.: Une condition géométrique pour l'inégalité de Harnack. J. Math. Pures Appl. (9) **64**(3), 237–256 (1985)
15. Franchi, B., Lanconelli, E.: Une métrique associée à une classe d'opérateurs elliptiques dégénérés. In: Conference on Linear Partial and Pseudodifferential Operators (Torino 1982), Rend. Sem. Mat. Univ. Politec. Torino 1983, Special Issue, 105–114 (1984)
16. Franchi, B., Lanconelli, E.: An embedding theorem for Sobolev spaces related to nonsmooth vector fields and Harnack inequality. Comm. Partial Differ. Equ. **9**(13), 1237–1264 (1984)
17. Franchi, B., Lu, G., Wheeden, R.L.: A relationship between Poincaré-type inequalities and representation formulas in spaces of homogeneous type. Int. Math. Res. Not. 1, 1–14 (1996)
18. Garofalo, N., Nhieu, D.-M.: Isoperimetric and Sobolev inequalities for Carnot-Carathéodory spaces and the existence of minimal surfaces. Comm. Pure Appl. Math. **49**(10), 1081–1144 (1996)
19. Gromov, M.: Carnot-Carathéodory spaces seen from within. Sub-Riemannian geometry, pp. 79–323, Progr. Math., 144, Birkhäuser, Basel (1996)
20. Gromov, M.: Metric structures for Riemannian and Non Riemannian Spaces, Progress in Mathematics, vol. 152. Birkhauser Verlag, Boston (1999)
21. Hajlasz, P., Koskela, P.: Sobolev met Poincaré. Mem. Am. Math. Soc. **145**(688) (2000)
22. Jerison, D.: The Poincaré inequality for vector fields satisfying Hörmander's condition. Duke Math. J. **53**(2), 503–523 (1986)
23. Jerison, D., Sánchez-Calle, A.: Estimates for the heat kernel for a sum of squares of vector fields. Indiana Univ. Math. J. **35**(4), 835–854 (1986)
24. Jurdjevic, V.: Geometric control theory. Cambridge Studies in Advanced Mathematics, vol. 52. Cambridge University Press, Cambridge (1997)
25. Kestin, J. (ed.): The Second Law of Thermodynamics. Dowden, Hutchinson & Ross, Inc. Stroudsburg, Pennsylvania (1976)
26. Kusuoka, S., Stroock, D.: Applications of the Malliavin calculus. III. J. Fac. Sci. Univ. Tokyo Sect. IA Math. **34**(2), 391–442 (1987)
27. Kusuoka, S., Stroock, D.: Long time estimates for the heat kernel associated with a uniformly subelliptic symmetric second order operator. Ann. Math. (2) **127**(1), 165–189 (1988)

28. Montgomery, R.: A tour of subriemannian geometries, their geodesics and applications. Mathematical Surveys and Monographs, vol. 91. American Mathematical Society, Providence (2002)

29. Murray, R.M., Sastry, S.S.: Nonholonomic motion planning: steering using sinusoids. IEEE Trans. Automat. Control **38**(5), 700–716 (1993)

30. Nagel, A., Stein, E.M., Wainger, S.: Balls and metrics defined by vector fields. I. Basic properties. Acta Math. **155**(1–2), 103–147 (1985)

31. Rashevski, P.K.: Any two points of a totally nonholonomic space may be connected by an admissible line. Uch. Zap. Ped. Inst. im. Liebknechta, Ser. Phys. Mat. **2**, 83–94 (1938)

32. Rothschild, L.P., Stein, E.M.: Hypoelliptic differential operators and nilpotent groups. Acta Math. **137**(3–4), 247–320 (1976)

33. Saloff-Coste, L.: A note on Poincaré, Sobolev, and Harnack inequalities. Internat. Math. Res. Notices , no. 2, 27–38 (1992)

34. Saloff-Coste, L., Stroock, D.W.: Opérateurs uniformément sous-elliptiques sur les groupes de Lie. J. Funct. Anal. **98**(1), 97–121 (1991)

35. Sánchez-Calle, A.: Fundamental solutions and geometry of the sum of squares of vector fields. Invent. Math. **78**(1), 143–160 (1984)

36. Sawyer, E.T., Wheeden, R.L.: Hölder continuity of weak solutions to subelliptic equations with rough coefficients. Mem. Am. Math. Soc. **180**(847) (2006)

37. Varopoulos, N.Th., Saloff-Coste, L., Coulhon, T.: Analysis and geometry on groups. Cambridge Tracts in Mathematics, vol. 100. Cambridge University Press, Cambridge (1992)

38. Varopoulos, N.Th.: Analysis on nilpotent groups. J. Funct. Anal. **66**(3), 406–431 (1986)

39. Varopoulos, N.Th.: Théorie du potentiel sur les groupes nilpotents. C. R. Acad. Sci. Paris Sér. I Math. **301**(5), 143–144 (1985)

Chapter 5
Beyond Hörmander's Operators

In this last chapter I want to discuss some developments which have taken place in the study of Hörmander's operators and related topics since the 1990s. As we will see, most of these developments have *extended* the class of operators under study, passing from classical Hörmander's operators to operators "structured on Hörmander's vector fields", in various senses, or operators also containing nonsmooth ingredients. In the end, most of these new kinds of operators are no longer hypoelliptic, but still share with classical Hörmander's operators several deep properties. I will try to illustrate the circle of ideas around some of these classes of operators.

I will not try, on the other hand, to make any account of the contemporary research about the *geometry of vector fields* (subriemannian geometry, geometric measure theory in subriemannian context, analysis in metric spaces and so on), another very active field of research which would require a separate focus.

More than in the previous sections, the choice of the topics covered here also reflects my personal tastes and research field.

5.1 Kolmogorov–Fokker–Planck Equations with Linear Drift

5.1.1 The Class of Operators Introduced by Lanconelli–Polidoro

We start describing some ideas contained in the paper by:

Lanconelli-Polidoro, 1994, Rend. Sem. Mat. Univ. Politec. Torino [50], which opened a new direction of research.

As we have seen in Sect. 2.1.4, the Kolmogorov equation corresponding to a system of stochastic o.d.e.'s reads as follows:

$$\sum_{i,j=1}^{N} a_{ij}(x,t)\, \partial^2_{x_i x_j} u + \sum_{k=1}^{N} b_k(x,t)\, \partial_{x_k} u - \partial_t u = 0, \qquad (5.1)$$

where (a_{ij}) is a symmetric nonnegative matrix. The paper [50] contains a deep study of the class of operators (5.1) having *constant matrix* $A = (a_{ij})$ and *linear drift*

M. Bramanti, *An Invitation to Hypoelliptic Operators and Hörmander's Vector Fields*, SpringerBriefs in Mathematics, DOI: 10.1007/978-3-319-02087-7_5, © The Author(s) 2014

$$b_k(x,t) = \sum_{j=1}^{N} b_{kj} x_j$$

with $B = (b_{kj})$ constant, i.e. operators of the form

$$L = \sum_{i,j=1}^{N} a_{ij} \partial_{x_i x_j} + \sum_{k,j=1}^{N} b_{kj} x_j \partial_{x_k} - \partial_t. \tag{5.2}$$

As we will see in the next sections, assuming the a_{ij} constant is just a starting point which will be generalized in subsequent researches. The particular structure of the drift, instead, is characteristic of this line of research.

These operators had been briefly discussed in the introduction of Hörmander's paper [40] and had been studied by Kuptsov in several papers appeared since 1972 to 1982 (see [50] for precise references). Let us define the matrix:

$$C(t) = \int_0^t E(s) A E^T(s) \, ds, \quad \text{where } E(s) = \exp\left(-s B^T\right). \tag{5.3}$$

Then one can prove (see [50, Appendix]) that:

The operator (5.2) is hypoelliptic if and only if $C(t) > 0 \ \forall t > 0$.

Hence the nondegeneracy of a single $N \times N$ matrix characterizes hypoelliptic operators within the class (5.2), like the nondegeneracy of the matrix A characterizes the class of elliptic operators. Moreover, under this condition (which we will assume satisfied), the authors show that there exists a basis in $\mathbb{R}^N$ such that the constant matrices $A = (a_{ij})$, $B = (b_{ij})$ take the following form:

$$A = \begin{bmatrix} A_0 & 0 \\ 0 & 0 \end{bmatrix}, \tag{5.4}$$

with $A_0 = (a_{ij})_{i,j=1}^{p_0}$ $(p_0 \le N)$ symmetric and positive definite:

$$v |\xi|^2 \le \sum_{i,j=1}^{p_0} a_{ij} \xi_i \xi_j \le \frac{1}{v} |\xi|^2 \tag{5.5}$$

for all $\xi \in \mathbb{R}^{p_0}$, some positive constant v;

$$B = \begin{bmatrix} * & B_1 & 0 & \dots & 0 \\ * & * & B_2 & \dots & 0 \\ \vdots & \vdots & \vdots & \ddots & \vdots \\ * & * & * & \dots & B_r \\ * & * & * & \dots & * \end{bmatrix} \tag{5.6}$$

where B_j is a $p_{j-1} \times p_j$ block with rank p_j, $j = 1, 2, \ldots, r$, $p_0 \geq p_1 \geq \cdots \geq p_r \geq 1$ and $p_0 + p_1 + \cdots + p_r = N$, while the blocks denoted by $*$ are arbitrary. The next picture should give a visual idea of the structure of B:

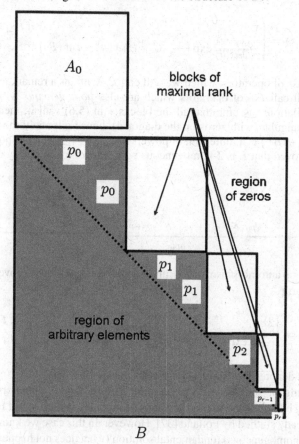

$$B$$

Note that the larger is the characteristic of the matrix A_0, the greater is our freedom in the choice of B. Let us assume that L is a hypoelliptic operator with the matrices A, B already written as (5.4) and (5.6). This operator turns out to be left invariant with respect to the following family of translations in $\mathbb{R}^{N+1}$, which form a Lie group:

$$(x, t) \circ (\xi, \tau) = (\xi + E(\tau)x, t + \tau),$$

where $E(\tau) = \exp\left(-\tau B^T\right)$. Note that the origin is the identity and

$$(\xi, \tau)^{-1} = (-E(-\tau)\xi, -\tau).$$

Also, an *explicit fundamental solution* for L can be constructed:

$$\Gamma(x,t;\xi,\tau) = \Gamma\left((\xi,\tau)^{-1} \circ (x,t)\right) = \Gamma(x - E(t-\tau)\xi, t - \tau)$$

where

$$\Gamma(x,t) = \begin{cases} 0 & \text{if } t \le 0 \\ \frac{(4\pi)^{-N/2}}{\sqrt{\det C(t)}} \exp\left(-\frac{1}{4}\langle C^{-1}(t)x, x\rangle - t\,\mathrm{tr}(B)\right) & \text{if } t > 0 \end{cases}.$$

The above class of operators, which we will call $\mathcal{L}$, contains a remarkable subclass, which we will call $\mathcal{L}_0$, of operators which are also *homogeneous* with respect to a family of dilations: assume that all the blocks $*$ in (5.6) vanish; then the matrix B is upper triangular (with zeros on the diagonal), hence nilpotent; the exponential $E(s) = \exp\left(-sB^T\right)$ is a finite sum of powers of $\left(-sB^T\right)$, hence is a polynomial in s. It can be proved that L is 2-homogeneous with respect to the family of dilations

$$D(\lambda)(x_1, x_2, \ldots, x_N, t)$$

$$= \left(\underbrace{\lambda x_1, \lambda x_2, \ldots, \lambda x_{p_0}}_{p_0}, \underbrace{\lambda^3 x_{p_0+1}, \ldots, \lambda^3 x_{p_0+p_1}}_{p_1}, \ldots, \underbrace{\lambda^{2r+1} x_{\ldots}, \ldots, \lambda^{2r+1} x_{\ldots}}_{p_r}, \lambda^2 t^2 \right)$$

which are group automorphisms. The fundamental solution can be proved to assume the simpler form

$$\Gamma(x,t) = \begin{cases} 0 & \text{if } t < 0 \\ \frac{c_N}{t^{Q/2}} \exp\left(-\frac{1}{4}\left\langle C^{-1}(1) D_0\left(\frac{1}{\sqrt{t}}\right) x, D_0\left(\frac{1}{\sqrt{t}}\right) x\right\rangle\right) & \text{if } t > 0 \end{cases}.$$

where $D_0(\lambda)$ denotes the dilations on the space variables only.

Before going on, some remarks are in order. The class $\mathcal{L}_0$ of operators considered above (having both translations and dilations) is included in the class of Hörmander's operators already studied by Folland [37]. However, in this case we know an *explicit form* for the (homogeneous) fundamental solution (what does not happen for general Folland's operators); moreover, this fundamental solution has the Gaussian-type form which we expect from a heat-type operator, and nevertheless the operator L is not just the heat-type operator corresponding to a sum of squares, that is

$$\sum_{i=1}^{q} X_i^2 - \partial_t,$$

but possesses a drift which involves both time and space variables. Finally, the larger class $\mathcal{L}$ of operators considered above (possessing translations but not dilations) represents an interesting intermediate case between Folland's setting and the general one.

Now, consider an operator $L \in \mathcal{L}$ (left invariant but nonhomogeneous); let L_0 be the similar operator which is obtained annihilating all the arbitrary elements $*$

in the matrix B. The authors' idea is that L_0, which is left invariant (with respect to a *different* group of translations!) and 2-homogeneous with respect to a suitable family of dilations, can be seen as the *principal part of* L. This idea is made precise in proving a comparison between the fundamental solutions of the two operators:

Theorem 80 *(See [50, Thm. 3.1]). Let L, L_0 be as above, and let Γ, Γ_0 be the fundamental solutions of L, L_0 (respectively) with pole $(0, 0)$. Then for every $b > 0$ there exists $a > 1$ such that*

$$\frac{1}{a}\Gamma_0(z) \leq \Gamma(z) \leq a\Gamma_0(z)$$

for any $z = (x, t) \in \mathbb{R}^{N+1}$ such that $\Gamma_0(z) \geq b$. Moreover $a = a(b) \to 1$ as $b \to +\infty$.

One could be surprised by the fact that the theorem requires that the fundamental solutions have pole at $(0, 0)$: after all, both the fundamental solutions are left invariant. The point is that the two groups of translations are different, so the approximation result holds true just near the origin. Actually, the authors exhibit a counterexample (see [50, Example 4.1]) showing that the quotient Γ/Γ_0 at a different pole can be unbounded.

It is worthwhile to work out a concrete example, to see the above ideas in action.

Example 81 *In $\mathbb{R}^3$, let*
$$L = \partial_{xx}^2 + x\partial_y + y\partial_x - \partial_t$$

which is an operator in $\mathcal{L}$, with

$$N = 2, \, p_0 = 1, \, A_0 = (1)$$

$$B = \begin{bmatrix} 0 & 1 \\ 1 & 0 \end{bmatrix} = B^T;$$

$$B^2 = I;$$

hence, for $k = 0, 1, 2, \ldots,$

$$B^{2k} = I, \, B^{2k+1} = B, \, and$$

$$E(s) = \exp\left(-sB^T\right) = \sum_{k=0}^{\infty} \frac{(-sB)^{2k}}{(2k)!} + \sum_{k=0}^{\infty} \frac{(-sB)^{2k+1}}{(2k+1)!}$$

$$= I\,Chs - B\,Shs = \begin{bmatrix} Chs & -Shs \\ -Shs & Chs \end{bmatrix}.$$

The operator L is left invariant with respect to the translations

$$(\xi_1, t_1) \circ (\xi_2, t_2) = (\xi_2 + E(t_2)\,\xi_1, t_1 + t_2),$$
$$(x_1, y_1, t_1) \circ (x_2, y_2, t_2) = (x_2 + x_1 Cht_2 - y_1 Sht_2, y_2 + y_1 Cht_2 - x_1 Sht_2, t_1 + t_2).$$

On the other hand, L is not homogeneous with respect to any family of dilations. (The nonexistence of a family of dilations adapted to the translation is evident since the operation $\circ$ is not polynomial; one can also realize that the vector field $x\partial_y + y\partial_x + \partial_t$ cannot be made 2-homogeneous with respect to any dilation). Let us compute the fundamental solution of L with pole at the origin.

$$E(s)\,AE^T(s) = \begin{bmatrix} Chs & -Shs \\ -Shs & Chs \end{bmatrix} \begin{bmatrix} 1 & 0 \\ 0 & 0 \end{bmatrix} \begin{bmatrix} Chs & -Shs \\ -Shs & Chs \end{bmatrix} = \begin{bmatrix} (Chs)^2 & -Shs\,Chs \\ -Shs\,Chs & (Shs)^2 \end{bmatrix}$$

$$C(t) = \int_0^t E(s)\,AE^T(s)\,ds = \frac{1}{2}\begin{bmatrix} Sht\,Cht + t & -(Sht)^2 \\ -(Sht)^2 & Sht\,Cht - t \end{bmatrix}$$

$$\det C(t) = \frac{1}{4}\left((Sht)^2 - t^2\right)$$

$$C^{-1}(t) = \frac{2}{(Sht)^2 - t^2}\begin{bmatrix} Sht\,Cht - t & (Sht)^2 \\ (Sht)^2 & Sht\,Cht + t \end{bmatrix}$$

$$\Gamma(x, y, t) = \begin{cases} 0 & \text{if } t \leq 0 \\ \dfrac{1}{2\pi\sqrt{(Sht)^2 - t^2}}\exp\left(-\dfrac{1}{2}\dfrac{(Sht\,Cht - t)x^2 + (Sht\,Cht + t)y^2 + 2xy(Sht)^2}{(Sht)^2 - t^2}\right) & \text{if } t > 0 \end{cases}$$

Now, let us consider the "principal part operator" obtained annihilating the element 1 in the lower left corner of the matrix B:

$$L_0 = \partial_{xx}^2 + x\partial_y - \partial_t.$$

This is an operator in $\mathcal{L}_0$, with

$$N = 2,\; p_0 = 1,\; A_0 = (1)$$

$$B_0 = \begin{bmatrix} 0 & 1 \\ 0 & 0 \end{bmatrix};$$

$$\left(B_0^T\right)^2 = 0;$$

hence

$$E_0(s) = \exp\left(-sB_0^T\right) = I - sB_0^T = \begin{bmatrix} 1 & 0 \\ -s & 1 \end{bmatrix}.$$

The operator L_0 is left invariant with respect to the translations

$$(\xi_1, t_1) * (\xi_2, t_2) = (\xi_2 + E_0(t_2)\xi_1, t_1 + t_2),$$
$$(x_1, y_1, t_1) * (x_2, y_2, t_2) = (x_1 + x_2, y_1 + y_2 - t_2 x_1, t_1 + t_2)$$

and is homogeneous with respect to the dilations

$$D(\lambda)(x, y, t) = \left(\lambda x, \lambda^3 y, \lambda^2 t\right).$$

Note that each component of the translation of L_0 can be seen as the principal part (in the sense of Taylor expansions) of the corresponding component of the translation of L, weighting the variables according to the dilations of L_0. Let us compute the fundamental solution Γ_0 of L_0 with pole at the origin.

$$E(s) A E^T(s) = \begin{bmatrix} 1 & 0 \\ -s & 1 \end{bmatrix} \begin{bmatrix} 1 & 0 \\ 0 & 0 \end{bmatrix} \begin{bmatrix} 1 & -s \\ 0 & 1 \end{bmatrix} = \begin{bmatrix} 1 & -s \\ -s & s^2 \end{bmatrix}$$

$$C(t) = \int_0^t E(s) A E^T(s)\, ds = \begin{bmatrix} t & -\frac{t^2}{2} \\ -\frac{t^2}{2} & \frac{t^3}{3} \end{bmatrix}$$

$$\det C(t) = \frac{t^4}{12}$$

$$C(t)^{-1} = \frac{12}{t^4} \begin{bmatrix} \frac{t^3}{3} & \frac{t^2}{2} \\ \frac{t^2}{2} & t \end{bmatrix} = \begin{bmatrix} \frac{4}{t} & \frac{6}{t^2} \\ \frac{6}{t^2} & \frac{12}{t^3} \end{bmatrix}$$

$$\Gamma_0(x, y, t) = \begin{cases} 0 & \text{if } t \le 0 \\ \frac{\sqrt{3}}{2\pi t^2} \exp\left(-\left(\frac{x^2}{t} + \frac{3y^2}{t^3} + \frac{3xy}{t^2}\right)\right) & \text{if } t > 0 \end{cases}.$$

It is a calculus exercise to check that, as $t \to 0^+$, the exponent and the factor of Γ_0 are the principal part of the exponent and the factor of Γ, weighting the variables (x, y, t) according to the homogeneities $(1, 3, 2)$.

As a consequence of the above comparison between the fundamental solutions of L and L_0, the authors are able to prove, for any operator of the class $\mathcal{L}$, an invariant Harnack inequality of parabolic type:

Theorem 82 *(See [50, Thm. 5.1']). Let $L \in \mathcal{L}$ and let Ω be an open subset of $\mathbb{R}^{N+1}$. Then there exist positive constants c, r_0 such that for any nonnegative solution u to $Lu = 0$ in Ω, every $z_0 \in \Omega$ and $0 < r \le r_0$ such that $H_r^-(z_0) \subset \Omega$ one has*

$$\sup_{H_r^-(z_0)} u \le cu(z_0)$$

where $H_r^-(z_0)$ is the base of a "parabolic cylinder", defined as follows:

$$H_r^-(z_0) = z_0 \circ H_r^- \text{ with } H_r^- = \left\{ (x, t) = \left(D_0(r) y, -r^2\right) : |y| \le 1 \right\}.$$

What makes the above theorem remarkable is the scaling invariance of this Harnack inequality, in contrast with the fact that operators in $\mathcal{L}$ do not possess a natural family of dilations.

5.1.2 Developments of the Theory of Homogeneous Operators of Lanconelli–Polidoro Type

The paper [50] has been the starting point of a line of research which is still lively. Operators in the class $\mathcal{L}_0$ have a very rich structure and possess an explicitly known fundamental solution of Gaussian type; they are the prototypes of interesting operators appearing in physics, finance, and other fields, so it has been natural to deepen their study. For instance, one can pass from the study of

$$L = \sum_{i,j=1}^{p_0} a_{ij} \partial_{x_i x_j} + \sum_{k,j=1}^{N} b_{kj} x_j \partial_{x_k} - \partial_t$$

with a_{ij} constant, to that of

$$L = \sum_{i,j=1}^{p_0} a_{ij}(x,t) \partial_{x_i x_j} + \sum_{k,j=1}^{N} b_{kj} x_j \partial_{x_k} - \partial_t \qquad (5.7)$$

or

$$L = \sum_{i,j=1}^{p_0} \partial_{x_i} \left(a_{ij}(x,t) \partial_{x_j} \right) + \sum_{k,j=1}^{N} b_{kj} x_j \partial_{x_k} - \partial_t. \qquad (5.8)$$

Note that as soon as the a_{ij}'s are not C^∞, these operators are no longer hypoelliptic. Here we start witnessing the study of operators which are somewhat *structured on Hörmander's vector fields*, but *do not have smooth coefficients*, a line of research which has been broadly pursued in the 2000's. For these operators one can try to prove a priori estimates in L^p or C^α spaces, for the second derivatives $\partial_{x_i x_j}$ with $i, j = 1, 2, \ldots, p_0$ or for the derivative $X_0 = \sum_{k,j=1}^{n} b_{kj} x_j \partial_{x_k} + \partial_t$, under suitable assumptions on the coefficients, study the Dirichlet problem, perform Moser's iterative method to prove various kinds of local estimates, and so on. Here we cannot give an account of the wealth of results obtained for this class of operators. A good survey can be found for instance in [49]. Just to give a sample of the works on this subject, let us say that nondivergence operators (5.7) have been studied for instance in [19, 47, 51, 67, 70]; divergence operators (5.8) have been studied for instance in [52, 68, 69, 71]; Moser's iterative method, and related issues, have been studied for instance in [64–66].

5.1.3 Developments of the Theory of Nonhomogeneous Operators of Lanconelli–Polidoro Type

Here we want to briefly sketch the content of a couple of more recent papers dealing with operators of Lanconelli-Polidoro type of the class $\mathcal{L}$ (that is, translation invariant but not homogeneous with respect to a family of dilations). As we will see, the study of this class of operators poses interesting geometric and real analysis problems, which have stimulated further research.

Di Francesco and Polidoro in [34] have proved Schauder-type estimates for non-divergence operators of type (5.7) assuming that the corresponding operator with constant a_{ij} belongs to the class $\mathcal{L}$, the matrix $\left(a_{ij}(x,t)\right)_{i,j=1}^{p_0}$ is symmetric and uniformly positive, and its entries a_{ij} are Hölder continuous, with respect to a distance which we will define in a moment. Let L be the operator under study, let K be the corresponding constant coefficient operator

$$ K = \sum_{i,j=1}^{p_0} a_{ij}\partial_{x_i x_j} + \sum_{k,j=1}^{n} b_{kj}x_j\partial_{x_k} - \partial_t . $$

(where the a_{ij} are any matrix in the same ellipticity class of the $a_{ij}(x,t)$), and let $K_0 \in \mathcal{L}_0$ be the principal part operator of K, in the sense of Lanconelli-Polidoro (see Sect. 5.1.1). The operator K is left invariant with respect to a group of translations $\circ$, which actually does not depend on the particular matrix a_{ij}, but just on its dimension and on the matrix $\left(b_{kj}\right)$; the operator K_0 is left invariant with respect to a different translation $*$ and is 2-homogeneous with respect to a family of dilations $D(\lambda)$; let $\|\cdot\|$ the homogeneous norm induced by these dilations; then Di Francesco–Polidoro introduce in $\mathbb{R}^{N+1}$ the function

$$ d(z,\zeta) = \left\| \zeta^{-1} \circ z \right\| . $$

Note that d looks like, but actually *is not* the standard quasidistance on a homogeneous group, because the norm $\|\cdot\|$ is relative to a homogeneous group, but the translation $\zeta^{-1} \circ z$ is relative to a different (nonhomogeneous) group. So this d is a "hybrid" object, which however can be proved to be, locally, a quasidistance, and actually is the right object to describe the geometry of the operator L. Namely, in [34] the authors are able to study the fundamental solution Γ of the operator K, showing that the kernel $\partial^2_{x_i x_j}\Gamma$ $(i, j = 1, 2, \ldots, p_0)$ satisfies with respect to the quasidistance d suitable pointwise bounds which allow them to prove $C^{2,\alpha}$ estimates on a domain of $\mathbb{R}^{N+1}$ for the variable coefficient operator L, when the $a_{ij}(x,t)$'s are C^{α}-Hölder continuous with respect to this d.

The operator K (with constant coefficients a_{ij}, possessing translations but not dilations) and its stationary counterpart

$$\mathcal{A}u = \sum_{i,j=1}^{p_0} a_{ij}\partial_{x_i x_j} + \sum_{k,j=1}^{n} b_{kj}x_j\partial_{x_k},$$

which is a hypoelliptic operator of Ornstein-Uhlenbeck type (and does not possess neither translations nor dilations), have been studied more recently by Bramanti, Cupini, Lanconelli and Priola [21], with the aim of proving *global L^p* estimates. Actually, the authors prove the following L^p bound on the strip $S \equiv \mathbb{R}^N \times [-1, 1] \subset \mathbb{R}^{N+1}$:

Theorem 83 *For every $p \in (1, \infty)$ there exists a constant $c > 0$ such that*

$$\left\|\partial^2_{x_i x_j}u\right\|_{L^p(S)} \le c\,\|Ku\|_{L^p(S)} \quad for\ i, j = 1, 2, \ldots, p_0,$$

for every $u \in C_0^\infty(S)$.

One of the problems to overcome, to get this result, is that the "hybrid" quasidistance d introduced in [34] to study this operator does not induce a structure of space of homogeneous type on the strip S. Namely, on the whole S the Lebesgue measure is doubling but the function $d(z, \zeta)$ is not a quasidistance, while on a bounded domain $\Omega \subset S$ the function d is a quasidistance but we cannot prove that the Lebesgue measure is doubling with respect to the restricted balls $\Omega \cap B_r$. This forces us to look for an original approach to singular integral theory. Actually, we have revised the *singular integral theory in nondoubling spaces*, a theory started around 2000 with the work of Tolsa, Nazarov-Treil-Volberg, and have extended it to cover the present case (see Bramanti [7]).

Applying the estimate in Theorem 83 to a function $u(x, t) = U(x)T(t)$ with T fixed cutoff function, the following *global L^p* bound for the corresponding stationary operator easily follows:

Theorem 84 *For every $p \in (1, \infty)$ there exists a constant $c > 0$, depending on p, N, p_0, the matrix B and the ellipticity constant of the matrix (a_{ij}) such that for every $u \in C_0^\infty(\mathbb{R}^N)$ one has:*

$$\left\|\partial^2_{x_i x_j}u\right\|_{L^p(\mathbb{R}^N)} \le c\left\{\|\mathcal{A}u\|_{L^p(\mathbb{R}^N)} + \|u\|_{L^p(\mathbb{R}^N)}\right\} \quad for\ i, j = 1, 2, \ldots, p_0$$

$$\tag{5.9}$$

$$\|Y_0 u\|_{L^p(\mathbb{R}^N)} \le c\left\{\|\mathcal{A}u\|_{L^p(\mathbb{R}^N)} + \|u\|_{L^p(\mathbb{R}^N)}\right\}. \tag{5.10}$$

Recall that for a general Hörmander's operator which does not possess neither translations nor dilations, by Rothschild-Stein's theory (see Section 3.4) we know that *local L^p* estimates on $\partial^2_{x_i x_j}u$ hold. Theorem 84 seems to be the first case of L^p estimates on the whole $\mathbb{R}^N$ proved for Hörmander's operators with no underlying group structure.

The previous results have been extended by the same authors to the case of Ornstein-Uhlenbeck operators $\mathcal{A}$ with variable continuous coefficients a_{ij}, in [20].

5.2 Nonlinear Equations Coming from the Theory of Several Complex Variables

5.2.1 Regularity Theory for the Levi Equation and the Study of "Nonlinear Vector Fields"

Another line of research related to Hörmander' s vector fields which in the last 15 years has given interesting results deals with the *Levi equation*, a nonlinear equation studied in the theory of several complex variables. Here we particularly refer to a series of papers on this topic by Citti, Montanari, Lanconelli and a few coauthors since the late 1990s (we will quote some of them in the following).

Let Ω be an open subset of $\mathbb{R}^3$ and $u : \Omega \to \mathbb{R}$ be a C^2 function. The *Levi curvature* of the graph of u (x, y, t), introduced by E. E. Levi in 1909 to characterize the domains of holomorphy in $\mathbb{C}^2$, is defined, at the point $(\xi, u\,(\xi))$ $(\xi = (x, y, t))$ as the number

$$k\,(\xi, u) = \frac{\left(1 + u_t^2\right)^{1/2}}{\left(1 + a^2 + b^2\right)^{3/2}} \mathcal{L}u \tag{5.11}$$

with

$$\mathcal{L}u = u_{xx} + u_{yy} + 2au_{xt} + 2bu_{yt} + \left(a^2 + b^2\right) u_{tt}$$

where a, b are actually functions of the gradient of u; namely,

$$a = \frac{u_y - u_x u_t}{1 + u_t^2}; \quad b = \frac{-u_x - u_y u_t}{1 + u_t^2}.$$

If $k\,(\xi)$ is an assigned function in Ω, (5.11) is the *equation of the prescribed Levi curvature*, and is quasilinear. If $u \in C^2\,(\Omega)$ is a solution to (5.11), then u can be seen as a solution to the linear equation

$$\mathcal{L}u = k\,(\xi) \frac{\left(1 + a^2 + b^2\right)^{3/2}}{\left(1 + u_t^2\right)^{1/2}} \tag{5.12}$$

where the right hand side and the coefficients a, b, c at the left hand side are known continuous functions (if k is at least continuous). Note, however, that for any function u the matrix of the coefficients of this linear operator,

$$\begin{bmatrix} 1 & 0 & a \\ 0 & 1 & b \\ a & b & (a^2 + b^2) \end{bmatrix}$$

has zero determinant, so the equation is degenerate, and the standard theory of quasilinear elliptic equations does not apply. The main theme of the papers we are describing is the proof of C^∞ smoothness of a solution to this equation, under suitable assumptions on the assigned curvature and the minimal regularity of the solution itself.

The starting idea contained in the papers by Citti [29] and [30][1] is that of defining the vector fields (depending on the solution u):

$$X = \left(1, 0, \frac{u_y - u_x u_t}{1 + u_t^2}\right), \quad Y = \left(0, 1, -\frac{u_x + u_y u_t}{1 + u_t^2}\right).$$

Then $\mathcal{L}$ can be formally written as

$$\mathcal{L}u = X^2 u + Y^2 u - c(u)\,\partial_t u$$

with

$$c(u) = X\left(\frac{u_y - u_x u_t}{1 + u_t^2}\right) - Y\left(\frac{u_x + u_y u_t}{1 + u_t^2}\right).$$

Moreover,

$$[X, Y] = -\frac{\mathcal{L}u}{1 + u_t^2}\,\partial_t. \tag{5.13}$$

Comparing (5.12) with (5.13) one sees that if the curvature k is everywhere nonzero, then X, Y satisfy Hörmander's condition. Note, however, that if the solution u is initially assumed just C^2 (or with any limited degree of smoothness), the vector fields X, Y are not smooth, and $\mathcal{L}$ cannot be seen as a hypoelliptic operator. The main result proved in [30] is the following:

Theorem 85 *(See [30, Thm. 1.1.]) Let $u \in C^{2,\alpha}(\Omega)$ be a solution to (5.12) for some $\alpha > 1/2$, and assume that $k = k(\xi) \neq 0$ for every $\xi \in \Omega$ and $k \in C^\infty(\Omega)$; then $u \in C^\infty(\Omega)$.*

This theorem can be seen as a regularity result for a solution to an equation built with *nonsmooth Hörmander's vector fields* (we could also say *nonlinear vector fields*). Note that under the initially assumed regularity $u \in C^{2,\alpha}(\Omega)$, the vector fields X, Y have $C^{1,\alpha}(\Omega)$ coefficients, and the equation is satisfied in $C^{0,\alpha}$. The idea is that of introducing the smooth approximating vector fields obtained replacing the coefficients of X, Y with their first order Taylor polynomials, and exploit existing

---

[1] The first of the two papers actually deals with a slightly simplified equation, but already contains the main ideas exploited in the second one to handle the Levi equation.

techniques of Hörmander's operators to handle these smooth vector fields. In doing so, the knowledge of the explicit form of these vector fields X, Y is of great help.

This result has been extended in several directions in subsequent papers. Just to quote a few of them: Citti and Montanari [32] have extended the result of [30] to the Levi equation in $\mathbb{R}^{2n+1}$; Citti, Lanconelli and Montanari [31] have proved the smoothness of any Lipschitz continuous solution to the Levi equation in $\mathbb{R}^3$, with nonvanishing assigned Levi curvature; this clearly requires first of all a suitable notion of generalized solution to the equation; Montanari [55] studied the regularity of solutions to the corresponding evolution equation (that is, regularity of real hypersurfaces evolving by Levi curvature). Further references can be found in the quoted papers, and checking the subsequent production of the same authors.

The main point that we want to stress in these papers is that they represent a motivation for and an example of study of *nonsmooth Hörmander's vector fields*, a topic which we will discuss again in Sect. 5.4.

5.2.2 Levi–Monge–Ampère Equations and Nonvariational Operators Structured on Hörmander's Vector Fields

A line of research related to the one we have just discussed is the study of the *Levi-Monge-Ampère equation*, a fully nonlinear degenerate elliptic equation which can be usefully analyzed in terms of Hörmander's vector fields. This has been studied by several authors. A couple of papers which are representative of the issues we are interested in are: Montanari and Lanconelli [56], and Montanari and Lascialfari [57].

Without going into details, we are interested in discussing how these researches have motivated the introduction of a new class of *nonvariational operators structured on Hörmander's vector fields*.[2]

"Many problems in geometric theory of several complex variables lead to fully nonlinear second order equations, whose linearizations are nonvariational operators of Hörmander type

$$L = \sum_{i,j=1}^{q} a_{ij}(x) X_i X_j. \tag{5.14}$$

Here we would like to present one of these problems whose source goes back to some papers by Bedford, Gaveau, Słodkowsky and Tomassini, see [2, 76, 77].

Let M be a real hypersurface, embedded in the Euclidean complex space $\mathbb{C}^{n+1}$. The Levi form of M at a point $p \in M$ is a Hermitian form on the complex tangent space whose eigenvalues $\lambda_1(p), \ldots, \lambda_n(p)$ determine in the directions of each corresponding eigenvector a kind of "principal curvature". Then, given a generalized symmetric function s, in the sense of Caffarelli, Nirenberg and Spruck [25], one can define the s-Levi curvature of M at p, as follows:

---

[2] What follows in this paragraph is extracted from the introduction of [13].

$$S_p(M) = s(\lambda_1(p), \ldots, \lambda_n(p)).$$

When M is the graph of a function u and one imposes that its s-Levi curvature is equal to a given function, one obtains a second order fully nonlinear partial differential equation, which can be seen as the pseudoconvex counterpart of the usual fully nonlinear elliptic equations of Hessian type, as studied e.g. in [25]. In linearized form, the equations of this new class can be written as (see [56, equation (34) p.324])

$$\mathcal{L}u \equiv \sum_{i,j=1}^{2n} a_{ij}\left(Du, D^2u\right) X_i X_j u = K(x, u, Du) \text{ in } \mathbb{R}^{2n+1} \qquad (5.15)$$

where:

the X_j's are first order differential operators, with coefficients depending on the gradient of u, which form a real basis for the complex tangent space to the graph of u;

the matrix $\{a_{ij}\}$ depends on the function s;

K is a prescribed function.

It has to be noticed that $\mathcal{L}$ only involves $2n$ derivatives, while it lives in a space of dimension $2n + 1$. Then, $\mathcal{L}$ is never elliptic, on any reasonable class of functions. However, the operator $\mathcal{L}$, when restricted to the set of strictly s-pseudoconvex functions, becomes "elliptic" along the $2n$ linearly independent directions given by the X_i's, while the missing one can be recovered by a commutation. Precisely,

$$\dim\left(\text{span}\left\{X_j, [X_i, X_j], i, j = 1, \ldots, 2n\right\}\right) = 2n + 1$$

at any point (see [56, equation (36) p. 324]). This is a Hörmander-type rank condition of step 2.

The parabolic counterpart of (5.15), i.e. equation

$$\partial_t u(t, x) = \mathcal{L}u(t, x) \text{ for } t \in \mathbb{R}, \ x \in \mathbb{R}^{2n+1} \qquad (5.16)$$

arises studying the evolution by s-Levi curvature of a real hypersurface of $\mathbb{C}^{n+1}$ (see [41, 55])."

5.3 Nonvariational Operators Structured on Hörmander's Vector Fields

We now want to discuss the line of research, developed in the 2000s, devoted to the study of "nonvariational operators" of the kind

$$Lu = \sum_{i,j=1}^{q} a_{ij}(x) X_i X_j \tag{5.17}$$

where $X_1, X_2, \ldots, X_q$ are Hörmander's vector fields defined in some domain $\Omega \subset \mathbb{R}^n$ ($n > q$) and the matrix $\{a_{ij}\}_{i,j=1}^{q}$ is symmetric and uniformly positive on $\mathbb{R}^q$ with (at least) bounded measurable coefficients:

$$\mu |\xi|^2 \leq \sum_{i,j=1}^{q} a_{ij}(x) \xi_i \xi_j \leq \mu^{-1} |\xi|^2 \tag{5.18}$$

for any $\xi \in \mathbb{R}^q$, a.e. $x \in \Omega$, some positive constant μ. Further regularity properties on the coefficients a_{ij} are imposed depending on the specific problem under study. In any case the a_{ij}'s are not assumed C^∞, hence the operator L is not hypoelliptic, but is a degenerate elliptic operator, structured on Hörmander's vector fields.

Related operators which have been investigated are the evolution counterpart of (5.17), namely

$$Lu = \sum_{i,j=1}^{q} a_{ij}(x,t) X_i X_j - \partial_t \tag{5.19}$$

and the more general operator with drift

$$Lu = \sum_{i,j=1}^{q} a_{ij}(x) X_i X_j + X_0 \tag{5.20}$$

where now $X_0, X_1, \ldots, X_q$ are Hörmander's operators in a domain $\Omega \subset \mathbb{R}^n$, $n > q+1$. Note that the class of operators (5.20) is (much) more general than (5.19); in particular, it contains the class (5.7) of nonvariational Kolmogorov-Fokker-Planck operators of Lanconelli-Polidoro type. Apart from the different structure of the three classes (5.17), (5.19), (5.20), a basic difference in the assumptions one can make on the vector fields consists in assuming the presence of an underlying homogeneous group or considering general Hörmander's vector fields. Here below we will survey the main results obtained so far on this subject.

5.3.1 L^p Estimates for Nonvariational Operators Structured on Hörmander's Vector Fields

Operators of type (5.17) appeared for the first time in the paper by Xu [78] in the context of Hölder estimates for a particular class of quasilinear equations. After that, the study of these operators intensified in the 2000s.

Bramanti and Brandolini in the two papers [8, 9] proved a priori estimates in Sobolev spaces induced by the vector fields. The first paper [8] considers general operators (5.20) but assumes the existence of a homogeneous group; the second one, [9] deals with general Hörmander's vector fields, but without drift (that is, operators (5.17)). The basic estimate proved in both cases reads as follows:

$$\|u\|_{S^{2,p}(\Omega')} \le c \left\{ \|Lu\|_{L^p(\Omega)} + \|u\|_{L^p(\Omega)} \right\}$$

for any $p \in (1, \infty)$, $\Omega' \Subset \Omega$, where

$$\|u\|_{S^{2,p}} = \|u\|_{L^p} + \sum_{j=1}^{q} \|X_j u\|_{L^p} + \sum_{i,j=1}^{q} \|X_i X_j u\|_{L^p} \; (+ \|X_0 u\|_{L^p}) \quad (5.21)$$

where the drift is present in case (5.20). Besides condition (5.18), the coefficients a_{ij} are assumed in the class VMO ("vanishing mean oscillation") with respect to the distance induced by the vector fields. To explain and motivate this assumption, we have to make a digression about the theory of nonvariational *elliptic* operators.

A classical result by Agmon, Douglis and Nirenberg [1] states that, for a given uniformly elliptic operator in nondivergence form with uniformly continuous coefficients,

$$Lu = a_{ij}(x) u_{x_i x_j}$$

one has the following L^p-estimates for every $p \in (1, \infty)$, on a bounded smooth domain Ω of $\mathbb{R}^n$:

$$\left\| u_{x_i x_j} \right\|_{L^p(\Omega)} \le c \left\{ \|Lu\|_{L^p(\Omega)} + \|u\|_{L^p(\Omega)} \right\}.$$

Classical counterexamples (see e.g. [53, Chap. 1]) show that the above estimate is generally false if the coefficients a_{ij} are merely L^∞. Miranda [54] proved that the same estimates hold if the coefficients belong to the Sobolev space $W^{1,n} \cap L^\infty$, a space which also contains discontinuous functions. A further remarkable extension of the classical result is due to Chiarenza, Frasca and Longo [27, 28] who proved that the continuity assumption can be replaced by the weaker condition $a_{ij} \in VMO \cap L^\infty$, where VMO is the space of *vanishing mean oscillation functions* (first introduced by Sarason in [74]), a sort of uniform continuity in integral sense. Namely, one asks that

$$\eta(r) \equiv \max_{i,j} \eta_{a_{ij}}(r) \to 0 \text{ as } r \to 0,$$

where, for any locally integrable function f, we set

$$\eta_f(r) = \sup_x \sup_{\rho < r} \left(\frac{1}{|B_\rho(x)|} \int_{B_r(x)} |f(y) - f_{B_\rho}| \, dy \right).$$

Since $W^{1,n}(\Omega) \subset VMO(\Omega)$, Miranda's result is contained in those proved by Chiarenza-Frasca-Longo. Let us sketch the argument followed by these authors. They start freezing the variable coefficients at some point x_0, writing a representation formula for u and then for $u_{x_i x_j}$ in terms of the frozen operator $L_0 u$; this formula involves the fundamental solution $\Gamma(x_0, \cdot)$ of L_0, which can be explicitly written. Next, one "unfreezes" the coefficients $a_{ij}(x_0)$, getting a representation formula for $u_{x_i x_j}$ which involves singular integrals acting on Lu,

$$T_{ij}(Lu)(x) = PV \int k_{ij}(x, x - y) Lu(y) dy$$

and commutators of singular integrals with the multiplication by the coefficients a_{ij}, acting on the second derivatives $u_{x_i x_j}$, that is

$$[T_{ij}, a_{hk}] u_{x_i x_j}(x) = PV \int k_{ij}(x, x - y) [a_{hk}(x) - a_{hk}(y)] u_{x_h x_k}(y) dy.$$

$$(5.22)$$

Taking L^p norms of both sides of the representation formula and exploiting L^p boundedness of the singular integrals T_{ij} one gets

$$\left\| u_{x_i x_j} \right\|_{L^p} \leq c \left\{ \|Lu\|_{L^p} + \sum_{h,k=1}^{n} \left\| [T_{ij}, a_{hk}] u_{x_h x_k} \right\|_{L^p} \right\}. \qquad (5.23)$$

Now a key ingredient comes in, namely a theorem by Coifman, Rochberg and Weiss [33] stating that the L^p norm of the commutator (5.22) can be bounded by

$$c \|a_{hk}\|_{BMO} \|u_{x_h x_k}\|_{L^p}.$$

If u is supported in a small ball, the BMO seminorm of the a_{hk}'s on this ball can be made small, thanks to the VMO assumption on the coefficients; this allows us to adsorb the term $\|u_{x_h x_k}\|_{L^p}$ to the left hand side of (5.23), and conclude the proof of the L^p bound.

This general strategy appears generalizable to any context where an explicit fundamental solution is known for the frozen operator, and a commutator theorem holds, with respect to the metric and balls induced by the operator. This program has been successfully carried out for nonvariational parabolic operators by Bramanti and Cerutti [18], 1993, and for operators of Lanconelli-Polidoro type of the class $\mathcal{L}_0$ (see Sect. 5.1.1) by Bramanti, Cerutti and Manfredini [19]. In the last case, the necessary extension of the commutator theorem to the context of spaces of homogeneous type has been given by Bramanti and Cerutti [17].

The paper [8] represents the first case when the same strategy has been successfully implemented notwithstanding the lacking of an explicit formula for the fundamental solution of the frozen operator: namely, given an operator of type (5.20) on a group, the corresponding frozen operator

$$Lu = \sum_{i,j=1}^{q} a_{ij}(x_0) X_i X_j + X_0$$

is a hypoelliptic, left invariant, homogeneous operator of Folland's type, admitting a homogeneous fundamental solution $\Gamma(x_0, \cdot)$ which, however, is not explicitly known. Nevertheless, we are able to prove, by an indirect argument, a uniform estimate of the kind

$$\sup_{x_0 \in \Omega} \sup_{|u|=1} \left| \partial_u^\alpha \Gamma(x_0; u) \right| \le c_\alpha \qquad (5.24)$$

for any multiindex α, with some constant depending on α, the homogeneous group and the ellipticity constant of the matrix $\{a_{ij}\}$. In some sense, one can say that the operators

$$Lu = \sum_{i,j=1}^{q} a_{ij}(x_0) X_i X_j + X_0$$

are proved to be "uniformly hypoelliptic" with respect to the frozen point x_0. The uniform bound (5.24) is necessary in order to reduce the study of singular integrals with kernels $X_i X_j \Gamma\left(u; v^{-1} \circ u\right)$ to singular integrals of convolution type.

Another novelty which is introduced in [8] is the use of interpolation inequalities for Sobolev norms defined by vector fields, to get the estimates (5.21) without first order derivatives at the right-hand side (what happened in [72]). Estimates (5.21) are also extended to higher order.

In [9] analogous results are established for general Hörmander's vector fields, but without a drift. Besides using the same machinery employed in [8], it is necessary here to exploit and adapt much of the material of [72]. The reason for not considering the drift, here, is that in the paper by Rothschild and Stein [72] the proofs are written in detail only in the case $X_0 = 0$, hence proving L^p estimates for nonvariational operators structured on Hörmander's vector fields with drift requires first of all a detailed revision and completion of some parts of [72], plus the solution of several new problems. This has actually been done, much more recently, by Bramanti and Zhu [23]. Let us mention here just one of these new problems which need to be solved to handle the drift case, which is related to the geometry of vector fields. Recall that the doubling property proved for metric balls by Nagel-Stein-Wainger (see Sect. 4.2) is of local type: for every $\Omega' \Subset \Omega$ there exist constants $c, r_0 > 0$ such that

$$|B(x, 2r)| \le c |B(x, r)| \quad \text{for any } x \in \Omega', r \le r_0. \qquad (5.25)$$

On the other hand, in order to apply in Ω' the existing abstract machinery of singular integrals in doubling spaces, we would need to know that

$$|B(x, 2r) \cap \Omega'| \le c |B(x, r) \cap \Omega'| \quad \text{for any } x \in \Omega', r \le r_0. \qquad (5.26)$$

Now, this global doubling (in Ω') (5.26) can be actually derived from its local version (5.25) as soon as Ω' is, say, a metric ball, provided the control distance satisfies a kind of "segment property", which means that given two points x_1, x_2 at distance R, for any $r < R$ and $\varepsilon > 0$ we can find another point x' at distance $\leq r$ from x_1 and $\leq R - r + \varepsilon$ from x_2. This fact has been proved for instance in [10, Lemma 4.2]. Now, the point is that this segment property is known to be true for the control distance when the drift is lacking, but is false in presence of a drift. Actually, a proof of (5.26) for the control distance of a system of vector fields with a drift of weight two has not been given yet. Therefore in [23] we had to exploit a different singular integral machinery, especially built for locally doubling spaces (see [24]).

Let us end this section with a brief account of other kinds of a priori estimates on the second order derivatives with respect to vector fields which have been proved through the years about nonvariational operators structured on Hörmander's vector fields. A first natural kind of estimates is in Hölder spaces $C_X^{2,\alpha}$. Here, for some $\alpha \in (0, 1]$

$$\|u\|_{C_X^{2,\alpha}(\Omega)} = \sum_{i,j=1}^{q} \|X_i X_j u\|_{C_X^{\alpha}(\Omega)} + \sum_{j=0}^{q} \|X_j u\|_{C_X^{\alpha}(\Omega)} + \|u\|_{C_X^{\alpha}(\Omega)}$$

where

$$\|u\|_{C_X^{\alpha}(\Omega)} = \|u\|_{C^0(\overline{\Omega})} + [u]_{C_X^{\alpha}(\Omega)} \text{ and }$$

$$[u]_{C_X^{\alpha}(\Omega)} = \sup_{x,y \in \Omega, x \neq y} \frac{|u(x) - u(y)|}{d_X(x, y)^{\alpha}}.$$

Estimates of the kind

$$\|u\|_{C_X^{2,\alpha}(\Omega')} \leq c \left\{ \|Lu\|_{C_X^{\alpha}(\Omega)} + \|u\|_{C^0(\overline{\Omega})} \right\}$$

for $\Omega' \Subset \Omega$ have been proved in [11] for operators (5.19), in [39] for operators (5.20) but in homogeneous groups, and in [23] for (5.20) and general Hörmander's vector fields. We also mention the paper [26], where a certain kind of "pointwise Schauder estimates" are proved for operators (5.17) on groups.

A different kind of a priori estimates are those of BMO type. This means that, under slightly stronger assumptions on the coefficients a_{ij} than those which allow to prove L^p estimates on $X_i X_j u$, one wants to bound also the BMO seminorm of $X_i X_j u$, in terms of the BMO seminorm of Lu plus the L^p norm of Lu and u itself. Without quoting the precise results, which should require some extra notation, we just mention the paper [10] where estimates of BMO type on $X_i X_j u$ are proved for operators (5.17) and general Hörmander's vector fields, assuming a suitable modulus of continuity on the coefficients a_{ij}, and [22], where estimates of BMO type on $X_i X_j u$ are proved for operators (5.17) on homogeneous groups, assuming the a_{ij} in $L^{\infty} \cap VLMO$, which also allows some kind of discontinuities.

5.3.2 Gaussian Estimates for Nonvariational Operators Structured on Hörmander's Vector Fields

A different approach to these classes of nonvariational operators is carried out by Bonfiglioli-Lanconelli-Uguzzoni in a series of papers appeared in the years 2002–2007 (see [3–6]) who, also motivated by the researches about fully nonlinear Levi-Monge-Ampère equations that we have described in Sect. 5.2.2, consider the class of evolution equations (5.19), with homogeneous left invariant vector fields on a Carnot group and Hölder continuous coefficients a_{ij}. The aim of the authors is to build a heat kernel for the variable coefficient evolution operator, prove that this kernel and its derivatives satisfy sharp Gaussian estimates, and use these results to prove an invariant parabolic Harnack inequality, following the approach which for nonvariational parabolic operators was first laid down by Fabes and Stroock [35] exploiting ideas by Nash [63].

The Gaussian bounds proved by the authors read as follows:

$$\frac{c_1}{t^{Q/2}} e^{-c_1 \|y^{-1} \circ x\|^2 / t} \le h(t, x, y) \le \frac{c_2}{t^{Q/2}} e^{-\|y^{-1} \circ x\|^2 / c_2 t}$$

$$\left| X_x^I X_y^J \partial_t^i h(t, x, y) \right| \le \frac{c}{t^{\frac{Q}{2} + i + \frac{|I| + |J|}{2}}} e^{-\|y^{-1} \circ x\|^2 / ct}$$

for any multiindices I, J, any nonnegative integer i. In turn, the construction of the heat kernel for the operator with Hölder continuous coefficients follows by a clever implementation of the classical Levi parametrix method, exploiting sharp Gaussian bounds for the heat kernel of the corresponding constant coefficient operator and for the difference of heat kernels corresponding to two different constant coefficient operators in the same ellipticity class of the coefficients a_{ij}. The whole research project is developed in the following steps: in [3] these uniform Gaussian bounds for heat kernels corresponding to constant coefficient operators are proved; in [4] the heat kernel for the operator with variable Hölder continuous coefficients is built, and Gaussian bounds for this kernel are proved; in [6] the "parabolic" Harnack inequality is proved. The first step deals with a problem which has some similarity with the proof of uniform estimates for fundamental solutions corresponding to frozen operators, which are proved in [8]; however, the nature of the bounds which are required in this case needs a completely different approach, which exploits the construction of a suitable diffeomorphism transforming any operator having the (constant) coefficient matrix $\{a_{ij}\}$ in a prescribed ellipticity class into another one fixed once and for all. This analysis is performed in still another paper (which therefore is logically the first of the series), namely [5].

It is natural to ask whether the previous results can be extended to general Hörmander's vector fields. This extension has been actually carried out in another research project by Bramanti, Brandolini, Lanconelli and Uguzzoni [13], see also [12] and [46]. Again, Harnack inequality is proved by Gaussian estimates on the heat kernel, and the heat kernel for the variable coefficient operator is built by the Levi parametrix method, exploiting uniform Gaussian bounds for the heat kernels of the frozen operator. However, the proof of these starting uniform bounds is not performed like in the case of groups, but actually exploits the results proved on groups, plus a good number of different techniques. In this case the desired Gaussian bounds assume the following form:

$$\frac{e^{-cd(x,y)^2/(t-s)}}{c\left|B\left(x,\sqrt{t-s}\right)\right|} \le h\left(t,x;s,y\right) \le c\frac{e^{-d(x,y)^2/c(t-s)}}{\left|B\left(x,\sqrt{t-s}\right)\right|}$$

$$\left|X_i h\left(t,x;s,y\right)\right| \le \frac{c}{\sqrt{t-s}}\frac{e^{-d(x,y)^2/c(t-s)}}{\left|B\left(x,\sqrt{t-s}\right)\right|} \tag{5.27}$$

$$\left|X_i X_j h\left(t,x;s,y\right)\right| + \left|\partial_t h\left(t,x;s,y\right)\right| \le \frac{c}{t-s}\frac{e^{-d(x,y)^2/c(t-s)}}{\left|B\left(x,\sqrt{t-s}\right)\right|}$$

where $x, y \in \mathbb{R}^n$, $0 < t - s < T$ and $|B(x, r)|$ denotes Lebesgue measure of the d-ball $B(x, r)$. The constant c in these estimates depends on the coefficients a_{ij} only through their Hölder moduli of continuity and the ellipticity constant.

Among the "auxiliary results" which are required in [13], there are also local a priori estimates of Schauder type for equations (5.19). These, which clearly have an independent interest, have been established, for general Hörmander's vector fields, by Bramanti and Brandolini [11] exploiting again Rothschild-Stein's "lifting and approximation" technique and the singular integral approach in the space of lifted variables already used in [9] to get L^p estimates, plus a study of singular integrals on Hölder spaces over spaces of homogeneous type, performed in the same paper. Here I would like to touch just one interesting point of the proof. Once we have proved the desired C^α estimates in the space of lifted variables, that is

$$\|u\|_{C_{\widetilde{X}}^{2,\alpha}(\widetilde{B}_r)} \le c\left\{\|\widetilde{H}u\|_{C_{\widetilde{X}}^{\alpha}(\widetilde{B}_{2r})} + \|u\|_{L^\infty(\widetilde{B}_{2r})}\right\},$$

we would like to "project" this bound on the original space, to get the final result. However, differently from the L^p case, applying the previous estimate to functions only depending on the original variables is not enough to get the desired result, because there is not a trivial relation between the distance d_X induced by the original vector fields in $\mathbb{R}^n$ and the distance $d_{\widetilde{X}}$ induced by the lifted vector fields in $\mathbb{R}^{n+m}$ (recall the discussion of this point in Sect. 4.2.5). Instead, one has to prove in this context Campanato's *integral characterization* of Hölder continuous functions, and then exploit the relation between the *volume* of lifted and unlifted balls (see again Sect. 4.2.5, Theorem 77).

Let us end this section pointing out that local Schauder estimates have been extended to nonvariational operators with drift (5.20) by Bramanti and Zhu [23].

5.4 Nonsmooth Hörmander's Vector Fields

5.4.1 Motivation and History of the Problem

A very recent field of research is that of *nonsmooth* general Hörmander's vector fields. To motivate it, let us recall some known results which we have discussed in previous sections.[3]

> "Starting from Hörmander's theorem, many other important properties have been proved, in the last 40 years, both regarding systems of Hörmander's vector fields and the metric they induce, and regarding second order differential operators structured on Hörmander's vector fields. In the first group of results, we recall:

> - the doubling property of the Lebesgue measure with respect to the metric balls [62];
> - Poincaré's inequality with respect to the vector fields [42].

> In the second group of results, we recall:

> - the "subelliptic estimates" of $H^{\varepsilon,2}$ norm of u in terms of L^2 norms of Lu and u [45];
> - the "$W^{2,p}$ estimates", involving second order derivatives with respect to the vector fields X_i, in terms of L^p norms of Lu and u [37, 72];
> - estimates on the fundamental solution of L or $\partial_t - L$ (again [36, 43, 62, 73]).

> Now, it is fairly natural to ask whether part of the previous theory still holds for a family of vector fields having only a partial regularity. Here are just a few facts which suggest this question:

> (i) to check Hörmander's condition of step r one has to compute derivatives of order up to $r - 1$ of the coefficients of vector fields;
> (ii) the definition of distance induced by a system of vector fields makes sense as soon as the vector fields are, say, locally Lipschitz continuous (in this general case, however, the distance of two points could be infinite, and proving connectivity, studying the volume of metric balls, proving the doubling condition and so on are open problems);
> (iii) apart from Hörmander's theorem about hypoellipticity, which is meaningful in the context of operators with C^∞ coefficients, several important results about second order differential operators built on Hörmander's vector fields are stated in a form which makes sense also for vector fields with a limited regularity (e.g., Poincaré inequality, a priori estimates on $X_i X_j u$ in L^p or Hölder spaces, *etc.*)

Actually, several authors have studied "nonsmooth Hörmander vector fields". The existing results up to 2010 or so can be classified mainly in two groups:

1. Nonsmooth vector fields having a particular form:

   1.1. "Diagonal" vector fields, that is having the special form:

---

[3] The following paragraph is extracted from the Introduction of [15].

$$X_i = a_i(x) \partial_{x_i} \quad i = 1, 2, \dots, n, \text{ in } \mathbb{R}^n \text{ with}$$

a_i Lipschitz continuous and possibly vanishing, hence

$$L = \sum_{i=1}^{n} X_i^2 \text{ is degenerate.}$$

These have been studied in several papers by Franchi-Lanconelli (see Sect. 4.6) in the 1980s, then Franchi [38], Sawyer and Wheeden [75].

1.2. "Nonlinear vector fields" studied in the context of equations of Levi-type since 1996 by Citti, Montanari, Lanconelli (see Sect. 5.2.1).

1.3. Nonsmooth vector fields of step two, studied by Montanari and Morbidelli [59, 60].

2. "Axiomatic theories" of general Lipschitz continuous vector fields, and metrics induced by these. (Already discussed in Sect. 4.5). For instance, one assumes the validity of a connectivity theorem, a doubling condition for balls, a Poincaré inequality, and then proves several consequences. A number of papers have been devoted to these researches, for instance by Capogna, Danielli, Franchi, Gallot, Garofalo, Gutierrez, Lanconelli, Hajlasz, Koskela, Morbidelli, Nhieu, Serapioni, Serra Cassano, Wheeden...

Now, all the previous results either assume a particular structure of the vector fields, or assume axiomatically the validity of some important properties. Instead, here we are interested in discussing the theory of *general nonsmooth Hörmander's vector fields*. This means to consider a family of vector fields $X_1, X_2, \dots, X_q$ which satisfy Hörmander's condition at step r and possess the minimal number of derivatives which are sufficient to check Hörmander's condition. Typically, one can ask $C^{r-1,\alpha}$ coefficients with some $\alpha \in [0, 1]$, depending on the kind of result. This theory has been studied by Bramanti, Brandolini and Pedroni [15, 16], Bramanti, Brandolini, Manfredini and Pedroni [14], and Montanari, Morbidelli in [58, 61]. Some related results have been obtained also by Karmanova and Vodopianov [44].

Why the nonsmooth extension of these results is not straightforward? After all, one could think that, for the results which are intrinsically meaningful under the assumption of a limited regularity of the vector fields, perhaps the same proofs which have been designed for the smooth case still hold in the nonsmooth one. Let us give at least some reasons of why this is not true.

As we have seen in the discussion of the paper by Nagel, Stein and Wainger [62] (see Sect. 4.2.3), a key tool in the study of the volume of metric balls is the diffeomorphism

$$\Phi : \{u_I\}_{I \in B} \mapsto x = \exp\left(\sum_{I \in B} u_I X_{[I]} \right)(x_0)$$

where $\left\{ \left(X_{[I]}\right)_{x_0} \right\}_{I \in B}$ is a basis of $\mathbb{R}^n$ obtained choosing, at the point x_0, suitable commutators of the X_i's. Now, if the vector fields X_i belong to $C^{r-1,1}$, the commutators $X_{[I]}$ for $|I| \leq r$ will be in general only $C^{0,1}$, and the map Φ will have only a $C^{1,1}$ regularity. In particular, it will be impossible to represent the vector fields X_i in these coordinates and compute their commutators of weight $|I| \leq r$ after the change of coordinates. To make this possible, we should assume a starting regularity of the X_i's of order C^{2r} or so, which seems definitely too restrictive to build an interesting theory.

A second obstacle to the repetition in the nonsmooth context of the proofs which hold in the smooth case is the important use which is made in [62] of the Campbell-Hausdorff formula, a theorem about the exponential of the commutator of vector fields which in its standard form requires C^∞ smoothness of the vector fields. Clearly, this tool must be banned in the nonsmooth setting.

5.4.2 Some Results from the Theory of Nonsmooth Hörmander's Vector Fields and Operators

Here I want to describe first some results about the geometry of nonsmooth Hörmander's vector fields, obtained in [15, 16], and then some more recent findings about nonsmooth Hörmander's operators of second order, proved in [14], which are also based on the previous set of results.

Besides the nonsmoothness of the coefficients, one of the features of this theory is that of keeping explicitly into account the possible existence of a "drift" X_0 of weight two. The theory in [15] is divided into three parts. In the first one we make the following:

Assumption (A). For some integer $r \geq 2$ and bounded domain $\Omega \subset \mathbb{R}^n$ the vector fields $X_1, X_2, \ldots, X_q$ belong to $C^{r-1}(\overline{\Omega})$, while X_0 belongs to $C^{r-2}(\overline{\Omega})$. Moreover, the commutators of *weighted* step up to r (i.e., $X_1, \ldots, X_q$ have weight 1 and X_0 has weight 2) span $\mathbb{R}^n$ for any $x \in \Omega$.

We can define the distance d_X induced by the vector fields and their commutators, as Nagel-Stein-Wainger do:

Definition 86 *For any $\delta > 0$, let $C(\delta)$ be the class of absolutely continuous mappings $\varphi : [0, 1] \longrightarrow \Omega$ which satisfy*

$$\varphi'(t) = \sum_{|I| \leq r} a_I(t) \left(X_{[I]}\right)_{\varphi(t)} \quad a.e.$$

with $a_I : [0, 1] \to \mathbb{R}$ measurable functions,

$$|a_I(t)| \leq \delta^{|I|}.$$

Then define

$$d(x, y) = \inf \{\delta > 0 : \exists \varphi \in C(\delta) \text{ with } \varphi(0) = x, \varphi(1) = y\}.$$

The only difference with the distance discussed in Sect. 4.2.2 is that now the number $|I|$ is not the *length* but the *weight* of the multiindex I, with X_0 weighting 2.

Then, the distance d is proved to satisfy the same relation with the Euclidean distance which holds in the smooth case: for every $\Omega' \Subset \Omega$ there exist c_1, c_2 such that

$$c_1 |x - y| \leq d(x, y) \leq c_2 |x - y|^{1/r} \quad \forall x, y \in \Omega'.$$

Moreover, the local doubling condition holds for the metric balls, and the same estimate of volume of metric balls proved by Nagel, Stein, Wainger [62] in the smooth case holds (see Sect. 4.2.3 for the statements).

In the second part we assume that, besides Assumption (A), the following holds:
Assumption (B). The drift X_0 belongs to C^1 (for $r > 2$ this is implicit in the previous assumption).

Let us introduce also the following

Definition 87 (Control distance) $\forall \delta > 0$, *let $C_1(\delta)$ be the class of absolutely continuous curves $\varphi : [0, 1] \longrightarrow \Omega$ such that*

$$\varphi'(t) = \sum_{i=0}^{q} a_i(t)(X_i)_{\varphi(t)} \text{ a.e.,}$$

with $|a_0(t)| \leq \delta^2, |a_i(t)| \leq \delta$ for $i = 1, 2, \ldots, q$.

Set
$$d_1(x, y) = \inf \{\delta > 0 : \exists \varphi \in C_1(\delta) \text{ with } \varphi(0) = x, \varphi(1) = y\}.$$

Then, the distance d_1 is proved to be locally equivalent to d. In particular, the connectivity property holds for the vector fields $X_0, X_1, \ldots, X_q$. Moreover, the doubling condition and the estimate on the volume of metric balls still hold with respect to the control distance d_1.

In the third and last part of the paper, we prove a Poincaré's inequality.
Assumption (C). Let $X_0 \equiv 0$ and let the X_i's $(i = 1, 2, \ldots, q)$ belong to $C^{r-1,1}$.
Then:

Theorem 89 (Poincaré's inequality) *For every $\Omega' \Subset \Omega$ there exist $c, r_0 > 0, \lambda \geq 1$ such that for any d_1-ball $B_1 = B_1(x, \rho)$, with $\rho \leq r_0, x \in \Omega', u \in C^1(\overline{\lambda B})$, we have*

$$\int_B |u(y) - u_B| \, dy \leq c\rho \int_{\lambda B} \sqrt{\sum_{j=1}^{q} |X_j u(y)|^2} \, dy.$$

By the existing "axiomatic theories" of Carnot-Carathéodory spaces, the previous results in particular imply:

- a stronger Poincaré's inequality with exponents (p, p) for any $p \in [1, \infty)$ and the same ball B at both sides;
- a Sobolev inequality;
- De Giorgi-Nash-Moser theory for variational operators structured on nonsmooth Hörmander's vector fields

$$Lu \equiv \sum_{i,j=1}^{q} X_i^* \left(a_{ij}(x) X_j u \right)$$

with

$$\lambda |\xi|^2 \leq \sum_{i,j=1}^{q} a_{ij}(x) \xi_i \xi_j \leq \lambda^{-1} |\xi|^2,$$

that is Hölder continuity of local weak solutions to $Lu = 0$ and an invariant Harnack inequality (on metric balls) for positive solutions.

Let us now give some ideas of the strategies used to get the above results. In principle, there exist two possible strategies to prove in the nonsmooth case a result which is known in the smooth one:

(1) to approximate nonsmooth vector fields with suitable smooth vector fields, and try to deduce the nonsmooth result from the smooth one by approximation;
(2) to prove from scratch the nonsmooth results, building proofs which hold under weaker assumptions.

When the first strategy (approximation) is applicable, this is the simplest way; sometimes however it is necessary to follow the second one. Our theory is built with a mix of the two strategies, looking at every step for the simplest way. In Part 1 (see above) the approximation technique succeeds. More precisely: let $S_i^{x_0}$ be the smooth vector fields obtained taking the Taylor expansions of order $r - 1$ of the coefficients of the X_i's at x_0. Then we prove that the following inclusions hold between the metric balls related to these smooth fields and the original ones:

$$B_{S^{x_0}}(x_0, c_1\rho) \subset B_X(x_0, \rho) \subset B_{S^{x_0}}(x_0, c_2\rho).$$

This fact immediately implies the local doubling condition for B_X and the estimate on the volume of B_X, via the analogous results for the smooth case.

In Part 2 a careful study of connectivity properties and the control distance is performed. Connectivity is a kind of result which cannot be easily proved by approximation, so a direct, nonsmooth approach is needed here. We have to study the exponential maps of nonsmooth vector fields, and some compositions of these, the so-called "quasi-exponential maps" already used by Nagel-Stein-Wainger. The idea is the following. In the smooth case, we know that

$$\exp\left(-tX\right)\exp\left(-tY\right)\exp\left(tX\right)\exp\left(tY\right)(x_0) = \exp\left(t^2\left[Y, X\right]\right)(x_0) + o\left(t^2\right).$$

However note that, if the vector fields are differentiable only a finite number of times, then the left hand side will have one degree of regularity more than $\exp\left(t^2\left[Y, X\right]\right)(x_0)$. Therefore we can use iterated compositions of exponentials of the basic vector fields to approximate the exponentials of commutators of any order and doing so we can build in a neighborhood of x_0 a local diffeomorphism of class at least C^1 which approximately transforms a metric ball into a coordinate box. This result, the most technical part of the paper, is related both to the connectivity property and to the equivalence of the two distances d, d_1 induced by the vector fields.

The third part of the paper (proof of Poincaré's inequality) combines different strategies and results. Poincaré's inequality is established using a general approach developed by Lanconelli and Morbidelli [48] and consisting into proving a property which the authors call "representability of metric balls by X-controllable almost exponential maps". To apply this theory is necessary to use the properties of the control distance proved in Part 2, but in order to build this almost exponential map we also use the smooth approximating vector fields (studied in Part 1). Also, to prove that this map is X-controllable one has to make some "explicit" computations, which crucially exploit the assumption of Lipschitz continuity of the highest order derivatives of the coefficients of the X_i (recall that in Part 3 we assume $C^{r-1,1}$ vector fields). Finally, for technical reasons, checking the assumptions of the theory which allows to prove Poincaré's inequality is easier under the additional assumption that the vector fields are free. Hence we first prove the result under this additional assumption, and then reduce the general case to this particular one. Doing so requires the use of Rothschild-Stein lifting theorem, in the nonsmooth context. This is part of the results which have been established in [16]. Actually, this paper contains a nonsmooth analog of Rothschild-Stein's lifting and approximation theorems, and the basic properties of the "map theta" (see Sect. 3.4). Here the main difference between the smooth and nonsmooth results regards the properties of the diffeomorphism $\Theta_\eta\left(\xi\right)$, which in this context is smooth in ξ but just Hölder or Lipschitz continuous in η, what makes this theory much less flexible to use than in the smooth case.

Let us now pass to consider Hörmander's operators

$$L = \sum_{i=1}^{q} X_i^2 + X_0$$

built with nonsmooth Hörmander's vector fields, which have been recently studied in [14]. Having at our disposal the results of lifting and approximation proved in [16], one could think to repeat the general strategy used by Rothschild-Stein in [72] to get some local a priori estimates on $X_i X_j u$ in terms of Lu. However, the strong lack of symmetry of the map $\Theta_\eta\left(\xi\right)$ built in the nonsmooth context in [16] prevents us to do so. Instead, a more promising idea is the following. Let us lift the operator L to another nonsmooth operator

$$\widetilde{L} = \sum_{i=1}^{q} \widetilde{X}_i^2 + \widetilde{X}_0$$

living on a higher dimension space $\mathbb{R}^N \ni (x, h)$, which can be locally approximated by a left invariant operator of Folland type

$$\mathcal{L} = \sum_{i=1}^{q} Y_i^2 + Y_0$$

possessing a homogeneous left invariant fundamental solution Γ. Consider now the function $\Gamma\left(\Theta_\eta\left(\xi\right)\right)$ and, by saturation of the lifted variables h in $\xi = (x, h)$, k in $\eta = (y, k)$, define the new kernel

$$P(x, y) = \int_{\mathbb{R}^m} \left(\int_{\mathbb{R}^m} \Gamma\left(\Theta_{(y,k)}(x, h)\right) \varphi(h) \, dh \right) \varphi(k) \, dk.$$

This P is a good candidate to be a *local parametrix* for the operator L. Namely, in [14] we are able to implement in this context the Levi parametrix method, building a local fundamental solution γ for L. More precisely, asking $X_i \in C^{r,\alpha}$ and $X_0 \in C^{r-1,\alpha}$ for some $\alpha > 0$, we show the existence of a function $\gamma(x, y)$, defined in a neighborhood U of a fixed point and satisfying the following properties:

1. γ is continuous in the joint variables for $x \neq y$,
2. the following derivatives of γ (with respect to the x variable) exists and are continuous in the joint variables:

$$X_i \gamma(x, y), \, X_j X_i \gamma(x, y), \, X_0 \gamma(x, y);$$

3. $\gamma(\cdot, y)$ is a weak solution to $L\gamma(\cdot, y) = -\delta_y$, that is:

$$\int_U \gamma(x, y) L^* \psi(x) \, dx = -\psi(y)$$

for any $\psi \in C_0^\infty(U)$, $y \in U$;
4. γ is a solution to the equation:

$$L\gamma(\cdot, y) = 0 \text{ in } U \setminus \{y\}, \text{ for any } y \in U.$$

5. The following upper bounds hold:

$$|\gamma(x, y)| \le c \frac{d^2(x, y)}{|B(x, d(x, y))|};$$

$$|X_i \gamma (x, y)| \le c \frac{d(x, y)}{|B(x, d(x, y))|};$$

$$|X_j X_i \gamma (x, y)|, |X_0 \gamma (x, y)| \le c \frac{1}{|B(x, d(x, y))|}$$

$$|X_i X_j \gamma (x_1, y) - X_i X_j \gamma (x_2, y)| \le c \left(\frac{d(x_1, x_2)}{d(x_1, y)} \right)^{\alpha - \varepsilon} \frac{1}{|B(x_1, d(x_1, y))|}$$

$$|X_0 \gamma (x_1, y) - X_0 \gamma (x_2, y)| \le c \left(\frac{d(x_1, x_2)}{d(x_1, y)} \right)^{\alpha - \varepsilon} \frac{1}{|B(x_1, d(x_1, y))|}$$

for every $x_1, x_2, y \in U$ such that $d(x_1, y) \ge 2d(x_1, x_2)$, $\varepsilon \in (0, \alpha)$ and $i, j = 1, 2, \ldots, q$, with c depending on ε.

6. The operator L is locally solvable, in the following sense. For any $\beta \in (0, \alpha)$, $f \in C_X^\beta (U)$, the function

$$w(x) = - \int_U \gamma (x, y) f(y) \, dy$$

is a $C_{X,loc}^{2,\beta} (U)$ solution to the equation $Lw = f$ in U.

These results are based on the theory previously developed in [15, 16] and on suitable abstract theories of singular integrals. One of the main difficulties relies on the fact that the whole proof lives in a Carnot-Carathéodory space where the measure of the ball $B(x, r)$ does not behave like a fixed power of the radius, which makes geometrical estimates particularly delicate. We also have to take into account the fact that, due to presence of the drift, the doubling condition can be exploited only in its local form (as explained at the end of Sect. 5.3.1). Also, the nonsmooth character of the vector fields X_i makes necessary to develop a deep analysis of the dependence of the the map $\Theta_\eta (\xi)$ and the remainder vector fields R_i^η on the "bad" variable η.

References

1. Agmon, S., Douglis, A., Nirenberg, L.: Estimates near the boundary of solutions of elliptic partial differential equations under general boundary conditions. Part I, Comm. Pure Applied Math. **12**, 623–727 (1959); Part II, Comm. Pure Applied Math. 17(1), 35–92 (1964)
2. Bedford, E., Gaveau, B.: Hypersurfaces with bounded Levi form. Indiana Univ. Math. J. **27**(5), 867–873 (1978)
3. Bonfiglioli, A., Lanconelli, E., Uguzzoni, F.: Uniform Gaussian estimates of the fundamental solutions for heat operators on Carnot groups. Adv. Differ. Equ. **7**, 1153–1192 (2002)
4. Bonfiglioli, A., Lanconelli, E., Uguzzoni, F.: Fundamental solutions for non-divergence form operators on stratified groups. Trans. Am. Math. Soc. **356**(7), 2709–2737 (2004)
5. Bonfiglioli, A., Uguzzoni, F.: Families of diffeomorphic sub-Laplacians and free Carnot groups. Forum Math. **16**(3), 403–415 (2004)
6. Bonfiglioli, A., Uguzzoni, F.: Harnack inequality for non-divergence form operators on stratified groups. Trans. Am. Math. Soc. **359**, 2463–2481 (2007)

7. Bramanti, M.: Singular integrals in nonhomogeneous spaces: L^2 and L^p continuity from Hölder estimates. Revista Matematica Iberoamericana **26**(1), 347–366 (2010)
8. Bramanti, L., Brandolini, M.: L^p-estimates for uniformly hypoelliptic operators with discontinuous coefficients on homogeneous groups. Rend. Sem. Mat. dell'Univ. e del Politec. di Torino. **58**(4), 389–433 (2000)
9. Bramanti, M., Brandolini, L.: L^p-estimates for nonvariational hypoelliptic operators with VMO coefficients. Trans. Am. Math. Soc. **352**(2), 781–822 (2000)
10. Bramanti, M., Brandolini, L.: Estimates of BMO type for singular integrals on spaces of homogeneous type and applications to hypoelliptic PDEs. Rev. Mat. Iberoamericana **21**(2), 511–556 (2005)
11. Bramanti, M., Brandolini, L.: Schauder estimates for parabolic nondivergence operators of Hörmander type. J. Differ. Equ. **234**(1), 177–245 (2007)
12. Bramanti, M., Brandolini, L., Lanconelli, E., Uguzzoni, F.: Heat kernels for non-divergence operators of Hörmander type. Comptes rendus - Mathematique. **343**(7), 463–466 (2006)
13. Bramanti, M., Brandolini, L., Lanconelli, E., Uguzzoni, F.: Non-divergence equations structured on Hörmander vector fields: heat kernels and Harnack inequalities. Mem. AMS **204**(961), 1–136 (2010)
14. Bramanti, M., Brandolini, L., Manfredini, M., Pedroni, M.: Fundamental solutions and local solvability for nonsmooth Hörmander's operators. Submitted, Preprint (2013). http://arxiv.org/abs/1305.3398
15. Bramanti, M., Brandolini, L., Pedroni, M.: Basic properties of nonsmooth Hörmander's vector fields and Poincaré's inequality. Forum Mathematicum. **25**(4), 703–769 (2013)
16. Bramanti, M., Brandolini, L., Pedroni, M.: On the lifting and approximation theorem for nonsmooth vector fields. Indiana Univ. Math. J. **59**(6), 1889–1934 (2010)
17. Bramanti, M., Cerutti, M.C.: Commutators of singular integrals on homogeneous spaces. Boll. Un. Mat. Ital. B (7) **10**(4), 843–883 (1996)
18. Bramanti, M., Cerutti, M.C.: $W_p^{1,2}$-solvability for the Cauchy-Dirichlet problem for parabolic equations with VMO coefficients. Comm. Partial Differ. Equ. **18**(9–10), 1735–1763 (1993)
19. Bramanti, M., Cerutti, M.C., Manfredini, M.: L^p-estimates for some ultraparabolic operators with discontinuous coefficients. J. Math. Anal. Appl. **200**(2), 332–354 (1996)
20. Bramanti, M., Cupini, G., Lanconelli, E., Priola, E.: Global L^p estimates for degenerate Ornstein-Uhlenbeck operators with variable coefficients. Mathematische Nachrichten, **286**(11–12), 1087–1101 (2013)
21. Bramanti, M., Cupini, G., Lanconelli, E., Priola, E.: Global L^p estimates for degenerate Ornstein-Uhlenbeck operators. Mathematische Zeitschrift **266**(4), 789–816 (2010)
22. Bramanti, M., Fanciullo, M.S.: BMO estimates for nonvariational operators with discontinuous coefficients structured on Hörmander's vector fields on Carnot groups. Advances in Differential Equations **18**(9–10), 955–1004 (2013)
23. Bramanti, M., Zhu, M.: L^p and Schauder estimates for nonvariational operators structured on Hörmander vector fields with drift. Anal. Partial Differ. Equ. ArXiv: 1103.5116v1 26. (2011, to appear)
24. Bramanti, M., Zhu, M.: Local real analysis in locally homogeneous spaces. Manuscripta Math. **138**(3–4), 477–528 (2012)
25. Caffarelli, L., Nirenberg, L., Spruck, J.: The Dirichlet problem for nonlinear second-order elliptic equations. III. Functions of the eigenvalues of the Hessian. Acta Math. **155**(3–4), 261–301 (1985)
26. Capogna, L., Han, Q.: Pointwise Schauder estimates for second order linear equations in Carnot groups. Harmonic analysis at Mount Holyoke (South Hadley, MA, 2001), pp. 45–69, Contemp. Math., 320, Amer. Math. Soc., Providence, RI (2003)
27. Chiarenza, F., Frasca, M., Longo, P.: Interior $W^{2,p}$-estimates for nondivergence elliptic equations with discontinuous coefficients. Ricerche Mat. **40**, 149–168 (1991)
28. Chiarenza, F., Frasca, M., Longo, P.: $W^{2,p}$-solvability of the Dirichlet problem for non divergence elliptic equations with VMO coefficients. Trans. Am. Math. Soc. **336**(1), 841–853 (1993)

29. Citti, G.: C^∞ regularity of solutions of a quasilinear equation related to the Levi operator. Ann. Scuola Norm. Sup. Pisa Cl. Sci. (4) **23**(3), 483–529 (1996)
30. Citti, G.: C^∞ regularity of solutions of the Levi equation. Ann. Inst. H. Poincaré Anal. Non Linéaire **15**(4), 517–534 (1998)
31. Citti, G., Lanconelli, E., Montanari, A.: Smoothness of Lipchitz-continuous graphs with non-vanishing Levi curvature. Acta Math. **188**(1), 87–128 (2002)
32. Citti, G., Montanari, A.: C^∞ regularity of solutions of an equation of Levi's type in $\mathbb{R}^{2n+1}$. Ann. Mat. Pura Appl. (4) **180**(1), 27–58 (2001)
33. Coifman,R.R., Rochberg, G.: Weiss: Factorization theorems for Hardy spaces in several variables. Ann. Math. **103**, 611–635 (1976)
34. Di Francesco, M., Polidoro, S.: Schauder estimates, Harnack inequality and Gaussian lower bound for Kolmogorov-type operators in non-divergence form. Adv. Differ. Equ. **11**(11), 1261–1320 (2006)
35. Fabes, E.B., Stroock, D.W.: A new proof of Moser's parabolic Harnack inequality using the old ideas of Nash. Arch. Ration. Mech. Anal. **96**, 327–338 (1986)
36. Fefferman, C., Sánchez-Calle, A.: Fundamental solutions for second order subelliptic operators. Ann. Math. (2) **124**(2), 247–272 (1986)
37. Folland, G.B.: Subelliptic estimates and function spaces on nilpotent Lie groups. Ark. Mat. **13**(2), 161–207 (1975)
38. Franchi, B.: Weighted Sobolev-Poincaré inequalities and pointwise estimates for a class of degenerate elliptic equations. Trans. Am. Math. Soc. **327**, 125–158 (1991)
39. Gutiérrez, C.E., Lanconelli, E.: Schauder estimates for sub-elliptic equations. J. Evol. Equ. **9**(4), 707–726 (2009)
40. Hörmander, L.: Hypoelliptic second order differential equations. Acta Math. **119**, 147–171 (1967)
41. Huisken, G., Klingenberg, W.: Flow of real hypersurfaces by the trace of the Levi form. Math. Res. Lett. **6**(5–6), 645–661 (1999)
42. Jerison, D.: The Poincaré inequality for vector fields satisfying Hörmander's condition. Duke Math. J. **53**(2), 503–523 (1986)
43. Jerison, D., Sánchez-Calle, A.: Estimates for the heat kernel for a sum of squares of vector fields. Indiana Univ. Math. J. **35**(4), 835–854 (1986)
44. Karmanova, M., Vodopyanov, S.: Geometry of Carnot-Carathéodory spaces, differentiability, coarea and area formulas. Anal. Math. Phys. Trends Math., pp. 233–335 (2009)
45. Kohn, J.J.: Pseudo-differential operators and hypoellipticity. Partial differential equations (Proc. Sympos. Pure Math., vol. XXIII, Univ. California, Berkeley, Calif., 1971), pp. 61–69. American Mathematical Society, Providence (1973)
46. Lanconelli, E.: Heat Kernels in Subriemannian settings. In: Proceeding of the CIME Summer Course 2007 on "Geometric Analysis and PDEs". Springer Lectures Notes in Math. vol. 2009, pp. 35–61 (1977)
47. Lanconelli, E., Lascialfari, F.: A boundary value problem for a class of quasilinear operators of Fokker-Planck type. In: Proceedings of the Conference "Differential Equations". Ann. Univ. Ferrara Sez. VII (N.S.) 41 (1996), suppl., 65–84 (1997)
48. Lanconelli, E., Morbidelli, D.: On the Poincaré inequality for vector fields. Ark. Mat. **38**(2), 327–342 (2000)
49. Lanconelli, E., Pascucci, A., Polidoro, S.: Linear and nonlinear ultraparabolic equations of Kolmogorov type arising in diffusion theory and in finance. Nonlinear problems in mathematical physics and related topics, II, 243–265, Int. Math. Ser. (NY), 2. Kluwer/Plenum, New York (2002)
50. Lanconelli, E., Polidoro, S.: On a class of hypoelliptic evolution operators. Partial differential equations, II (Turin, 1993). Rend. Sem. Mat. Univ. Politec. Torino **52**(1), 29–63 (1994)
51. Manfredini, M.: The Dirichlet problem for a class of ultraparabolic equations. Adv. Differ. Equ. **2**(5), 831–866 (1997)
52. Manfredini, M., Polidoro, S.: Interior regularity for weak solutions of ultraparabolic equations in divergence form with discontinuous coefficients. Boll. Unione Mat. Ital. Sez. B Artic. Ric. Mat. (8) **1**(3), 651–675 (1998)

53. Maugeri, A., Palagachev, D.K., Softova, L.G.: Elliptic and parabolic equations with discontinuous coefficients. Math. Res. vol. 109, Wiley (2000)
54. Miranda, C.: Sulle equazioni ellittiche del secondo ordine di tipo non variazionale, a coefficienti discontinui. Ann. Mat. Pura Appl. (4) **63**, 353–386 (1963)
55. Montanari, A.: Real hypersurfaces evolving by Levi curvature: smooth regularity of solutions to the parabolic Levi equation. Comm. Partial Differ. Equ. **26**(9–10), 1633–1664 (2001)
56. Montanari, A., Lanconelli, E.: Pseudoconvex fully nonlinear partial differential operators: strong comparison theorems. J. Differ. Equ. **202**(2), 306–331 (2004)
57. Montanari, A., Lascialfari, F.: The Levi Monge-Ampère equation: smooth regularity of strictly Levi convex solutions. J. Geom. Anal. **14**(2), 331–353 (2004)
58. Montanari, A., Morbidelli, D.: A Frobenius-type theorem for singular Lipschitz distributions. J. Math. Anal. Appl. **399**(2), 692–700 (2013)
59. Montanari, A., Morbidelli, D.: Sobolev and Morrey estimates for non-smooth vector fields of step two. Z. Anal. Anwendungen **21**(1), 135–157 (2002)
60. Montanari, A., Morbidelli, D.: Balls defined by nonsmooth vector fields and the Poincaré inequality. Ann. Inst. Fourier (Grenoble) **54**(2), 431–452 (2004)
61. Montanari, A., Morbidelli, D.: Nonsmooth Hörmander vector fields and their control balls. Trans. Am. Math. Soc **364**, 2339–2375 (2012)
62. Nagel, A., Stein, E.M., Wainger, S.: Balls and metrics defined by vector fields. I. Basic properties. Acta Math. **155**(1–2), 103–147 (1985)
63. Nash, J.: Continuity of solutions of parabolic and elliptic equations. Am. J. Math. **80**, 931–954 (1958)
64. Pascucci, A., Polidoro, S.: A Gaussian upper bound for the fundamental solutions of a class of ultraparabolic equations. J. Math. Anal. Appl. **282**(1), 396–409 (2003)
65. Pascucci, A., Polidoro, S.: The Moser's iterative method for a class of ultraparabolic equations. Commun. Contemp. Math. **6**(3), 395–417 (2004)
66. Pascucci, A., Polidoro, S.: On the Harnack inequality for a class of hypoelliptic evolution equations. Trans. Am. Math. Soc. **356**(11), 4383–4394 (2004)
67. Polidoro, S.: On a class of ultraparabolic operators of Kolmogorov-Fokker-Planck type. Le Matematiche (Catania) **49**(1), 53–105 (1995)
68. Polidoro, S.: Uniqueness and representation theorems for solutions of Kolmogorov-Fokker-Planck equations. Rend. Mat. Appl. (7), 15 (1995), no. 4, 535–560 (1996)
69. Polidoro, S.: A global lower bound for the fundamental solution of Kolmogorov-Fokker-Planck equations. Arch. Ration. Mech. Anal. **137**(4), 321–340 (1997)
70. Polidoro, S., Ragusa, M.A.: Sobolev-Morrey spaces related to an ultraparabolic equation. Manuscripta Math. **96**(3), 371–392 (1998)
71. Polidoro, S., Ragusa, M.A.: Hölder regularity for solutions of ultraparabolic equations in divergence form. Potential Anal. **14**(4), 341–350 (2001)
72. Rothschild, L.P., Stein, E.M.: Hypoelliptic differential operators and nilpotent groups. Acta Math. **137**(3–4), 247–320 (1976)
73. Sánchez-Calle, A.: Fundamental solutions and geometry of the sum of squares of vector fields. Invent. Math. **78**(1), 143–160 (1984)
74. Sarason, D.: Functions of vanishing mean oscillations. Trans. Am. Math. Soc. **207**, 391–405 (1975)
75. Sawyer, E.T., Wheeden, R.L.: Hölder continuity of weak solutions to subelliptic equations with rough coefficients. Mem. Am. Math. Soc. **180**, 847 (2006)
76. Slodkowski, Z., Tomassini, G.: Weak solutions for the Levi equation and envelope of holomorphy. J. Funct. Anal. **101**(2), 392–407 (1991)
77. Tomassini, G.: Geometric properties of solutions of the Levi-equation. Ann. Mat. Pura Appl. (4) **152**, 331–344 (1988)
78. Xu, C.J.: Regularity for quasilinear second-order subelliptic equations. Comm. Pure Appl. Math. **45**(1), 77–96 (1992)